NUTS A
OF THE PAST

Other books by David Freeman Hawke

EVERYDAY LIFE IN EARLY AMERICA
JOHN D.
THOSE TREMENDOUS MOUNTAINS
FRANKLIN
IN THE MIDST OF A REVOLUTION
THE COLONIAL EXPERIENCE
BENJAMIN RUSH

NUTS & BOLTS OF THE PAST

A History
of American Technology
1776–1860

David Freeman Hawke

HARPER & ROW, PUBLISHERS, NEW YORK
GRAND RAPIDS, PHILADELPHIA, ST. LOUIS, SAN FRANCISCO
LONDON, SINGAPORE, SYDNEY, TOKYO, TORONTO

First PERENNIAL LIBRARY edition published 1989.

The Library of Congress has catalogued the hardcover edition as follows:

Hawke, David Freeman.
  Nuts and bolts of the past.

  Includes index.
  1. Technology—United States—History—18th century. 2. Technology—United States—History—19th century. 3. Inventions—United States—History—18th century. 4. Inventions—United States—History—19th century.
I. Title.
T21.H39   1988   609'.73   87-46145
ISBN 0-06-015901-4
ISBN 0-06-091605-2 (pbk.)

89 90 91 92 93 FG 10 9 8 7 6 5 4 3 2 1

Contents

Acknowledgments

A variety of fears strike the author who prepares a manuscript for publication: Has he misspelled the name of a central figure or misquoted an author? Are all quotations accurate? The list of queries to oneself go on and on. Another crucial worry is: Has someone been overlooked in the Acknowledgments who contributed substantially to the book?

Fortunately, a new tradition has been established in publishing that allows usually unsung ladies and gentlemen within the publisher's office to be recognized in the opening pages of a book. Some, like the designer and the copyeditor of this book, I have never met, yet I, as must every author, offer thanks for giving the best possible public appearance to a manuscript. Others—like my editor, Hugh Van Dusen; his assistant, Stephanie Gunning; and especially Carole Herbin, who traipsed about the East searching out illustrations—I know and have personally thanked, but a public acknowledgment of their work is called for and is now given.

Here, however, the largest fear enters. A book that has been so long in the making, as this one, owes debts to many that I may overlook. If any names are missing, the fault lies with my shadowy memory. I am, and have been for some fifteen years, indebted to my former colleagues at Lehman College, the City University of New York. Two, however, have been my guard of honor: Davis Ross, who knows more about American technology than I ever will, and Jacob Judd, who never asks for anything and always gives much.

Three friends—William Doll, Norton H. Morrison, and Ludwig T. Smith—have read the galleys of this book. Each, with grace and generosity, have made substantial emmendations. Thanks go to all of them.

Life must be lived forward, but it can only be understood backwards.

—Søren Kierkegaard

I am not digging into such things because I think the old ways are necessarily better than the new ways, but I think there may be some of the old ways that we would be wise to look into before all knowledge of them disappears from the earth—the knowledge, and the kind of thinking that lay behind it.

—Robertson Davies

Introduction

A Cautious Approach to a Difficult Subject—Some of the Pitfalls That Lie Ahead

One afternoon in the 1830s Sidney George Fisher, a young Philadelphia socialite, came upon a large and handsome house on the outskirts of the city. It sat on the brow of a cliff overlooking a "romantic"* view. Fisher had to admit it exemplified good taste. John Price Wetherill owned it, a fact that struck Fisher as incongruous. Wetherill, always alert to technological innovations, had built the family chemical business into a fortune. Fisher granted that he was intelligent, full of energy and enterprise, but he was also "the most filthy and slovenly man in town," and "his uncombed locks, unshaven face, and shabby attire give him great influence and popularity, particularly among the 'great unwashed.'" Naturally, Fisher's and Wetherill's paths never crossed socially, which was just as well; though they shared the same city, they lived in separate worlds and had little to say to each other. While Wetherill circulated among the great unwashed,

* The sources for all quotations are identified in the Notes and Bibliography, following the text.

1

Fisher spent his day reading English newspapers at the Athenaeum and playing billiards at his club. His hero was Ralph Waldo Emerson. He did not approve of change except when it added to his comfort; then he accepted as his due what those from Wetherill's world had to offer. His table setting gleamed at night under the glow of gas lights. When indoor plumbing became available, he had a water closet installed, also a bathtub with spigots that provided hot and cold water. He approved of the "hoisting machine" when it appeared in the Continental Hotel. "You enter a nicely furnished little room ten feet square," he said. "A man pulls a string and the room ascends with an easy motion. You can stop and get out at any story." He bought a sewing machine for his wife. "This ingenious little machine performs in an hour as much work as could be done with a needle in a day, and it is very pleasant employment to use it—many ladies become very fond of the occupation and prefer it to a piano." He enjoyed the luxury of a stateroom on the palatial steamboat that took him to Newport and Boston, and though it meant touching elbows with the "great unwashed," he took the railroad rather than the stage to New York.

Fisher accepted each convenience as it came along but showed no curiosity about the people who had created it. Later his son wrote successful biographies about several of America's founding fathers, William Penn and Benjamin Franklin, among others. The founding fathers of American technology appear nowhere in his works. So, for the most part, has it been with students of the past ever since. The people who declared America to be an independent nation have been honored, but those who did much to make that independence a reality have generally been overlooked. The Fishers' disdain for these men with dirty fingernails has until recently been passed down through the generations largely intact.

Even a casual probing into the nuts and bolts of the American past makes it clear that facts are not easy to come by. The modern scholar Eugene Ferguson, with his usual perception, comments on one difficulty:

The secrecy with which much of the machine building art was practiced in Europe during the period of its most fruitful development has put beyond our reach many of the answers to questions that we would ask. For example, who built the first effective metal planing machine? The 1817 machine of Richard Roberts, attributed and dated years later, exists. However reasonable the attribution and dating may be, the first published description of a metal planer was by Joseph Clement in the early 1830s. Meanwhile, there were many other skilled and enterprising craftsmen in whose secret rooms developments and innovations were being hammered out nearly in parallel.

The murky history of card-making machines suggests other pitfalls that loom ahead. Cards were used to comb snarls from wool and cotton fibers before they were spun into yarn or thread. They were made from stiff leather pads into which scores of bent wire teeth were inserted. By the time of the American Revolution, Nathan Sellers of Philadelphia had built a machine to pierce holes in the pads, another to cut the wire into uniform lengths, and still another to bend the wire teeth to the proper angle. But the cards still had to be assembled by hand. "Every tooth," Sellers's grandson George Escol Sellers said, "had to be separately picked up and by nimble fingers put into the holes pierced to receive them." Sellers put out the work to a roster of three thousand women and children, and at any time he kept between three hundred and four hundred of these people busy assembling cards.

Obviously, a machine that could automate this process would speed the work and reduce costs considerably. Sometime around 1780 Oliver Evans of Delaware invented such a machine; it inserted wire teeth into the pads at the rate of three thousand a minute. Automation had been achieved, so it seemed. Not so. Half a century later Evans's machine, or one like it, "was in use to some extent, but owing to the shape and largeness of the pierced holes in the leather and consequent looseness of the teeth, had not then supplanted the more costly hand process."

The story now gets murkier. The record states that in 1797 Amos Whittemore of New England invented a card-making machine so efficient that "he made redundant a Boston factory where 60 men and 2,000 children made cards by hand." We know that such a machine existed and was attributed to Whittemore, for Zachariah Allen in *The Science of Mechanics* described how it worked in 1829: It "draws off the card wire from the reel, cuts it off at a proper length for the teeth, bends it into the form of a staple, punctures the holes in the leather, and inserts the staples of wire into the punctures, and finally crooks the teeth to the desired form—performing all these operations with regularity without the assistance of the human hand to guide or direct it." Yet much about Whittemore and his machine are suspect.

The authoritative *Dictionary of American Biography* states that Whittemore was born in Cambridge, Massachusetts, but the English historian H. J. Habakkuk says he was an Englishman who came to America when he could not sell his invention at home. One American historian, Roger Burlingame, thinks he stole Evans's design, while another, William Weeden, thinks not. Regardless, it is unlikely his machine ended hand assembly, for it, like Evans's, produced cards that were "coarse and imperfect," according to a contemporary Englishman. But this judgment, too, must be taken with care; English bias against American products may warp it. The defects of Whittemore's machine, says the Englishman, were no hindrance to its acceptance in America because standards of quality were lower there than in England, so "coarse cards were more in requisition." Moreover, "the want of efficient card makers in America rendered a self-acting machine of value there, although not very perfect in operation." Another contemporary, George Escol Sellers again, adds to the confusion. He dismisses the importance of both Evans's and Whittemore's creations and credits the first efficient card-making machine to John Brandt, a blacksmith from Lancaster, Pennsylvania. Brandt did not produce his invention until the late 1820s, nearly half a century after Evans's came on the market.

This muddy history of card-making machines only hints at the pitfalls that lie ahead. Contemporary biases aside, facts are difficult to pin down because little was done to preserve an accurate record. Partly the blame falls on the men who might have made the record—"much writing ill suits a mechanic," David Rittenhouse said—but the fault also lies with society, notably its historians. "One cannot blame the industrialist who dumped into the river apparently worthless documents," Siegfried Giedion has written. "The historians who did not succeed in awakening a feeling for the continuity of history are to blame."

A society preserves what it treasures, and America has not until recently treasured its technological past. No relics of Evans's or Whittemore's or Brandt's machines have come down to us. A buffalo robe brought back from the West by Lewis and Clark and displayed by Charles Willson Peale in his Philadelphia museum survives today in a Harvard museum, but the variety of mechanical models also exhibited by Peale—one of a steam dredge built by Oliver Evans, another of an iron bridge designed by Thomas Paine, and one of Matthias Baldwin's early locomotives—have all vanished.

Foreign visitors for the most part have done little to flesh out the record; few came before the mid-nineteenth century who were interested in the nuts and bolts of the American story. Michel Chevalier, a Frenchman, was contemptuous of English visitors who found in America ''a great deal that was bad and scarcely any thing good.'' Americans, he said, ''have a right to be tried by their peers, and it does not belong to the most complete aristocracy in Europe, the English aristocracy, to sit in judgment on a democracy.... A Yorkshire farmer or a Birmingham mechanic would certainly pass a very different judgment; they would probably be as exclusively disposed to praise as the most disdainful tourists have been to blame.'' Another Frenchman, Alexis de Tocqueville, toured the city hall of New York and the state house in Albany, but if in either town he stepped inside a machine shop, he left no account of the visit. He talked at length to politicians throughout the land, but never to mechanicians like Nathan Sellers. Foreigners came by the score to see the Lowell textile factories, but few if any went down to the machine shop to interview Paul Moody or, later, James Francis, who helped design and build the first efficient water turbine, a major technological leap into the future.

Frances Trollope left a full record of the social and cultural life of Cincinnati, but she never visited one of the city's famous slaughterhouses, where progenitors of the modern assembly line transformed hogs into pork chops; she spoke only of the smells that came from such houses, ''odours that I will not describe, and which I heartily hope my readers cannot imagine.'' Even Harriet Martineau, another Englishwoman made of sturdier stuff, told what went on in the slaughterhouses only ''as it was told to me.''

Fortunately, a more intrepid traveler, Captain Frederick Marryat, braved the smells and left a record of what he saw. "Cincinnati is the pork-shop of the Union; and in the autumnal and early winter months, the way they kill pigs here is, to use a Yankee phrase, *quite a caution,*" he wrote in the late 1830s.

> Almost all the hogs fed in the oak forests of Ohio, Kentucky, and western Virginia are driven into this city, and some establishments kill as many as fifteen hundred a day; at least so I am told. They are despatched in a way quite surprising, and a pig is killed upon the same principle as a pin is made—by division or, more properly speaking, by combination of labor. The hogs confined in a large room are driven into a smaller one; one man knocks them on the head with a sledge-hammer and then cuts their throats; two more pull away the carcass, when it is raised by two others, who tumble it into a tub of scalding water. His bristles are removed in about a minute and a half by another party, when the next duty is to fix a stretcher between his legs. It is then hoisted up by two other people, cut open, and disembowelled; and in three minutes and a half from the time that the hog was grunting in his obesity, he has only to get cold before he is again packed up, and reunited in a barrel to travel all over the world.

When Siegfried Giedion sought to illustrate this scene in his book *Mechanization Takes Command,* he was astonished to learn that he could turn up no sketch earlier than 1873, even with the help of local historians. "It was explained, not too convincingly," he writes, "that Cincinnati was at first ashamed to trace its wealth to pork packing. All the city's activities, its muscial life, for instance, can be accurately followed. But in tracing the first mechanization of the butcher's trade and the beginnings of the assembly line we have no foothold."

Any history of the machines that built America, to use Roger Burlingame's apt phrase, must not assume that an advance on one front routinely swept through society. It is generally accepted that two Americans invented the first workable airplane. "The United States, with its enormous spaces and scattered population, was made for the airplane," John Keegan has said. But the nation did not quickly exploit what the Wright brothers gave it. "The initiative soon passed elsewhere. By 1913, Russian aircraft were bigger, French faster, German stronger, than any that were being built in the United States. The Wrights had grown middle-aged; and a native American aircraft industry had not grown with them.''

Fernand Braudel has warned that in the history of technology ''there is no single onward movement, but many actions and reactions, many changes of gear. It is not a linear process.'' When America was still a land of farmers in the early nineteenth century, it needed a cheap, efficient plow to open up the land. Thomas Jefferson designed one on scientific principles. Jethro Wood produced one of cast iron with interchangeable parts, which Jefferson promoted in letters to friends. In 1828 James Fenimore Cooper wrote: ''I have seen more graceful, and convenient ploughs in positive use here than are probably to be found in the whole of Europe united. In this single fact may be traced the history of the character of the people and the germ of their future greatness.'' The American farmer who had cleared the stumps from his fields and could afford one rapidly took first to the iron and then to the steel plow; but the historian must note that acceptance was not a linear process. The Smithsonian Institution's Museum of American History today displays in the agricultural section a wooden plow that was in service in the 1870s; except for the iron tip, it differs little from the gnarled wooden plows used by the colonists in the late seventeenth century.

To avoid the most damaging pitfall of all—a rambling, aimless narrative—boundaries must be set, which calls for confronting the word ''technology.'' Leo Marx and Dirk Struik err when they say that Jacob Bigelow, a physician and Harvard lecturer, coined the word in 1829 with his book *Elements of Technology*. Actually, it has a long history—see the *Oxford English Dictionary*—and as David Noble remarks, Bigelow only brought it ''into general usage'' in America. Incidentally, historians of technology dismiss Bigelow's work as ''a pretty thin potion,'' yet Perry Miller's praise for the author and his effort to redress an unbalanced view of the past call for attention. Miller writes that ''it is indeed curious that the highly industrialized society of twentieth-century America can be bullied by humanistic professors into remembering Emerson's *Nature* of 1836 or even to cherishing the candlesticks and spinning wheels of our preindustrial past, and yet will not bother to salute in Bigelow a prophet more relevant to the later economy than either Emerson or Jefferson.''

The definition of technology favored by Miller and found in most dictionaries—''the application of science, especially to industrial or commercial objectives''—will not do. Science through the eighteenth century and much of the nineteenth only occasionally affected American technology. Fernand Braudel's definition—''human history in all its diversity''—is too broad to serve. Better is Edwin Layton's: ''a spectrum with ideas at one end and techniques and things at the other, with design as a middle term.'' Better still is David Billington's: ''Technology is the making of things that did not previously exist.'' But let us be more specific and coin a definition adapted to the pages that follow:

> Technology makes it possible to do something never done before (the airplane), to do mechanically something previously done manually (the sewing machine), or to do more effectively something previously done mechanically (the repeating rifle).

This definition may leave out much, but for an account of the nuts and bolts of the American past down to the Civil War it should keep us from wandering too far afield.

I

A New Nation Conceived:

1776–1820

The newly born United States was more successful than any other nation in assuming the attitude of mind required and in transferring any desired technology. How could this be? How could the custodians of an empty continent, far from the distant power centers of Europe and its workshops, move to take leadership in one line after another of mechanization and innovation?

—Brooke Hindle, *Emulation and Invention*

1

A Backward Glance

The early settlers came from various parts of England. All brought with them a desire to create again what they had left behind. This desire, they soon learned, presented problems. Often the dialects they spoke differed so much that they could barely communicate with one another. The daily lives of all had been shaped by deep-rooted customs that none wanted to abandon.

Marriage, for instance, of a young lady to a man from a different part of England raised difficulties unknown back home, where most couples came from similar backgrounds. She might have been reared in a thatched hut, he in a stone house with a slate roof. Her father might have been accustomed to the open-field system of farming, where he tended strips of land in common with other farmers of the community and along with them herded his cattle on the village green. The young man might have come from a region where farmers worked their own fenced-in plots of land when and as they wished. Both were no doubt Protestants, but she might have been a dissenting Puritan, he an Anglican.

Still, those who came, regardless of their region of England, brought with them much in common—eyeglasses, which nearly doubled the productive life of a craftsman; the mechanical clock, which few had much need for; the compass; the spinning wheel; the plow; gunpowder; the wheelbarrow; the skill to build water-wheels and windmills. The lathe, powered by a treadle, was their single machine tool. Tool kits varied little among craftsmen, be they cabinetmakers, shipwrights, coopers, or whatever. They con-

11

tained hatchets, saws (handsaws, ripsaws to cut with the grain, crosscut saws to work across the grain), chisels, adzes, hammers, mallets, augers, gimlets, bit braces, axes (felling axes for trees, broadaxes for squaring timbers). Most of the tools in size and shape dated back to the Middle Ages. Few new ones were to be added during colonial times, for as Rob Tarule has said, "to introduce a new tool may cause sufficient disruption of the total system of tool use to be actually more a hindrance than an aid."

Few of the settlers were innovative people but circumstances forced changes on them. Some twenty thousand immigrants arrived in Massachusetts Bay within a decade. In 1682 alone two thousand settlers flooded into the Delaware Valley, and another two thousand came the next year. The instant need for shelter led in particular locales to something akin to standardized housing, usually shaped by an itinerant master builder or carpenter. When a site had been chosen he passed out orders to the client—uprights must be of a certain thickness and length, rafters of another size, the clapboards of a particular shape. He then moved on to another site, returning to supervise the raising of the frame when his instructions had been carried out. As early as the 1720s a Philadelphia carpenter advertised for sale standardized window frames, painted and glazed, for immediate installation. Some years later another carpenter was selling timbers "of good quality and well seasoned" that "could be purchased ready for assembling" for houses twenty feet wide and forty-four feet deep.

By the end of the seventeenth century ethnic variety had complicated what had begun as English settlements. Dutch, Germans, Scandinavians, Scots, Scots-Irish, and a smattering of Jews brought over their own customs and traditions. Like tended to marry like and live among their own kind, preserving all they could of the culture they had left behind. The unique mixture of settlers led the predominant English to a slow, reluctant altering of old ways of doing things. From the Dutch many accepted the Dutch oven and the Dutch cellar, the yacht, the sled, and ice skating. Scandinavians brought the log cabin, a cheap, easy, and swift way to build a shelter, but until the nineteenth century and the westward movement few settlers beyond the Delaware Valley adopted it. The most extensive borrowings during the colonial period came from the Indians.

The Indians had developed a culture perfectly adapted to the wilderness, and despite an underlying contempt toward his way of life the white man drew much from it. From the Indians he learned how to girdle trees—that is, to cut a band of bark around the trunk—which deadened the limbs and let enough sunlight seep into the forest to plant a crop. Few settlers were patient enough to learn how to build a birchbark canoe, one of the most efficient and graceful crafts ever to glide along inland waterways, but they did accept the Indian dugout canoe. In a land where shoes were expensive and shoemakers few, the American farmer adopted Indian footwear—the moccasin—and buckskin clothing. He refused, however, to accept the Indians' greatest technological offering—the bow and arrow, the ideal weapon for life in the wilderness, drawn from the forest itself. It was easy to construct, easy to replace. The owner need not depend on others for his ammunition or on some specialist to repair it. It was accurate over three times the range of a musket, and in the time it took a settler to shoot and reload his gun, an Indian could send from eight to ten arrows toward his target. But the settlers refused even to experiment with the bow and arrow; and, ironically, the Indians' eagerness to own guns helped to enchain them, for only white men could repair them and provide ammunition.

Not all innovations in early America came from other cultures. The heavy, awkward ax brought from England had by the eighteenth century evolved into a nicely balanced, lighter ax that took a deeper bite with less effort. By the eve of the Revolution the European scythe's straight blade and handle had been altered into the graceful tool of today. A wooden frame or cradle attached to the blade let the reaper collect his swath into a neat sheaf. The settlers also invented a new way to fence their fields that satisfied needs unique to America. Englishmen bounded their plots with sturdy hedges or post-and-rail fences—verticle posts planted deep in the ground with crossbeams nailed to them. Americans could not wait for hedges to grow; the post-and-rail fence called for nails, a costly item, and it took too long to erect for a people in a hurry. So they invented the zigzag or snake fence, which piled crisscrossed split rails atop one another to form a string of giant Zs. It needed no nails to hold the rails in place, only a slanted stake at each interlacing corner. It could be built quickly.

The nearly equal weight of the bit or cutting edge and the poll or flat edge gave the American axe exceptional balance. "In contrast," write Hindle and Lubar, "the European axe had a longer and narrower bit and hardly any poll at all. This difference permitted the American axe to be swung straight and clean, without the slightest wavering. In addition, the wooden handle was given a length and curve precisely fitted to the height and swing of the axman. The result was remarkable. A practiced American axman could fell three times as many trees in the same time as a man using a European axe." *(Smithsonian Institution)*

A man handy with an ax and wedge could cut 150 to 200 rails ten feet long in a day, while a coworker could during the same day convert the rails into two hundred yards of fence. It was durable and easy to repair, and could be torn down and reassembled to enclose a newly opened field. Chesapeake planters invented the snake fence, but it soon spread throughout the colonies and was still in use well over a century later, when Abraham Lincoln became known as the "railsplitter." Europeans saw it as a monstrosity. It wasted a strip of about ten feet of tillable land wherever it zigged and zagged, and it used an extravagant amount of lumber, but it suited Americans, who had an abundance of wood and land.

They also had an abundance of fast-running streams that led down to the sea, which, tied to the abundance of wood, led to another early innovation—the sawmill. Wood in England by the early years of the seventeenth century had become a precious commodity. Forests had all but vanished into planks for ships (it took more than 2,500 trees to build an oceangoing vessel), into fireplaces and stoves, into charcoal for forges and furnaces. To get the most from every tree, skilled sawyers converted it into lumber with a thin-bladed ripsaw, a tedious process but one that left less than one-eighth behind as sawdust, compared to one-fifth of each log cut by the thick blade of a water-powered saw. (Americans handled all that debris in cavalier fashion. Some shoveled it into a neighboring swamp or marsh, but most dumped it into the stream that turned the waterwheel, making once-clear water turbid. The polluted stream soon drove away fish, and where once Indians and whites had annually harvested tons of herring and salmon, soon there were none.)

The early Americans lacked the skill to build sawmills; England at the time and for decades thereafter had none. Yet within four years after the first ships arrived in Chesapeake Bay they had one on the James River, built by German craftsmen especially imported for the job. The Dutch, accustomed to wind-driven mills, put one up soon after settling New Amsterdam. Four years after the settlement of Massachusetts Bay a sawmill appeared there, this time built by Danish experts; by the end of the seventeenth century the colony had seventy-two water-driven mills. Along the sparsely settled New Hampshire and Maine coasts some sixty mills were turning out hundreds of thousands of boards annually.

The proliferation of sawmills helps to debunk a myth about early American technology—that it emerged from a shortage of labor, especially skilled labor. Erecting a sawmill called for more than installing a mechanically powered blade. A variety of craftsmen were needed to build the mill's foundation and the shed that sheltered the blade, the dam and sluice that assured a steady flow of water, and above all the waterwheel. Craftsmen from the European Continent charted the way, but the early settlers, once they had studied the pattern, quickly learned to reproduce it on streams up and down the coast. They taught themselves to do what they had never done before because American circumstances forced them to be more flexible than they had been back home.

If a single word can characterize these people, it is *flexibility*. The failure of English guilds to cross the ocean and control the price and production of goods—guilds "were more apt to restrict innovation than encourage it, out of fear of allowing advantage to their more ingenious members," according to Anthony Wallace—helped to speed change. Benjamin Franklin's father came to Boston a guild member trained as a dyer of wool. He could not find enough work to keep the family alive, and so he switched to making soap, a product the town needed. England divided the medical profession into three branches—physician, surgeon, and apothecary. In America three became one, for the lack of business. The doctor, as he came to be called, blended his own prescriptions, amputated limbs, and served as a general practitioner. Similarly, a gunsmith who had specialized in constructing lock mechanisms for muskets branched out into other sidelines. Charles Willson Peale trained as a saddler but, Brooke Hindle remarks, "when saddlery proved insufficiently remunerative he moved through a succession of other trades with an ease that astonishes us today—and that would have been impossible in the more highly rigidly controlled environment of England." The need for bridles led him into silversmithing. From a villager's defective watch "he acquired knowledge of such machines" and soon after advertised himself as one who could repair them. John Fitch, a pioneer in building steamboats, began as a clockmaker and became a brass founder, later a silversmith, a cartographer, and an engraver. Later still, during the Revolution, he became a gunsmith. So the story went everywhere in early America. Paul Revere, silversmith, ended his life building boilers for Fulton's steamboat.

Why early America welcomed technology can be answered in a variety of ways. An American has suggested that the colonist "generally worked as an individual, without membership in a trade organization with others of the same craft." A Frenchman thought that the American "is a mechanic by nature," that he conforms easily "to new situations and circumstances; he is always ready to adopt new processes and implements or to change his occupation." An Englishman has said that the self-employed American brought "his own methods of doing a job better." None of these vague yet reasonable insights traces back to an institution that took a unique form in America—the farm—and to the type of farmer it produced.

It is hard to generalize about early American farmers. Judgments of them vary from Jefferson's idealization—"Those who labor in the earth are the chosen people of God, if ever He had a chosen people, whose breasts He has made His peculiar deposit for substantial and genuine virtue"—to the condemnation by a seventeenth-century Virginian who called them men who "sponge upon the blessings of a warm sun and fruitful soil, and almost grutch the pains of gathering in the bounties of the earth." This much, though, can be said with certainty: they differed within a few years of settlement from their English counterparts not only in what they were and in what they raised but in the way they lived as well. The relatively isolated farm not only affected social life and familial relations but also forced men to do for themselves what they had once depended on village craftsmen to do for them. Michel St. Jean de Crèvecoeur (called in the United States Hector St. John Crèvecoeur) described the American farmer as "a universal fabricator like Crusoe." What Crèvecoeur had to say about the eighteenth-century farmer holds for the man who came into being in the seventeenth century. "For most of us are skillful enough to use [tools] with some dexterity in mending and making whatever is wanted on the farm," he wrote. "Were we obliged to run to distant mechanics, who are half farmers

An early American plow with an iron tip. Such plows varied in size and shape with the talents of the farmers who made them. (*Smithsonian Institution*)

themselves, many days would elapse, and we should always be behind in our work.'' Crèvecoeur goes on: ''Does either his plough or his cart break, he runs to his tools; he repairs them as well as he can. Do they finally break down, with reluctance he undertakes to rebuild them, though he doubts his success. This is an occupation committed before to the mechanic of his neighborhood, but necessity gives him invention, teaches him to imitate, to recollect what he has seen. Somehow or another 'tis done.''

The flexibility of Crèvecoeur's farmer has endured to the present. Verlyn Klinkenborg records a conversation with his uncle Louie, a modern midwestern farmer faced with installing a new sickle blade on his windrower.

''Damn thing didn't fit,'' Louie said.

''What'd you do?''

''Fixed it.''

That terse explanation from Uncle Louie, a man of few words, leads Klinkenborg into an eloquent passage that describes the life of American farmers regardless of era:

> The Unexpected stalks a farm in big boots like a vagrant bent on havoc.
>
> Not every farmer is an inventor, but the good ones have the seeds of invention within them. Economy and efficiency move their relentless tinkering, and yet the real motive often seems to be aesthetic. The mind that first designed a cutter bar is not far different from the mind that can take the intractable steel of an outsized sickle blade and make it hum in the end. The question is how to reduce the simplicity that constitutes a problem (''It's simple; it's broke.'') to the greater simplicity that constitutes a solution.

The boy reared on a farm, then and now, finds himself employed in a machine shop. He brings no particular skills with him, but he does carry an open-minded attitude toward the work in hand, an attitude denied—in the past at least—an English craftsman constrained by traditions and standards imposed by his guild.

There was no shortage of workshops to attract lads bored with the lonely, monotonous life of farming and fascinated with machinery. Carl Bridenbaugh years ago disabused us of the long-accepted view that on the eve of the Revolution ''there was an

almost complete dearth of skilled artisans in the United States.''
Craftsmen found it hard to flourish in the South, where ''for the
want of towns, markets, and money,'' they were ''inexorably
drawn into planting and farming because they could not gain a
living from their trades.'' But they prospered everywhere in the
North. In Connecticut, for instance, lay ''the village precursors
of the great industries located today at Bridgeport, New Haven,
New Britain, Wallingford, and Meriden. Their line of descent is
clear and unbroken from the 1730s to the present and by the time
of the Revolution were deep-rooted.'' What Bridenbaugh says of
Connecticut holds true for all of New England, as well as for
Pennsylvania and New York, Delaware and New Jersey. But the
country as a whole was only dimly aware of what it had. Colonies,
and within the colonies hamlets, villages, and towns, lived lives
largely apart. It took a war for the new nation to get to know
itself.

2

The Spirit of '76

When Thomas Jefferson in 1776 spoke of "my country" he meant Virginia, and similarly John Adams meant Massachusetts. So it was with every man in the Second Continental Congress. Few of the fifty-five delegates, except for Benjamin Franklin and perhaps one or two others, knew much about the land that stretched beyond their own colony's borders. Samuel Adams had never been out of Massachusetts until he came to the congress. The outbreak of fighting at Lexington forced them all to expand their vision. "When fifty or sixty men have a constitution to form for a great empire," said John Adams, "at the same time they have a country of fifteen hundred miles extent to fortify, millions to arm and train, a naval power to begin, an extensive commerce to regulate, numerous tribes of Indians to negotiate with, a standing army of 27,000 men to raise, pay, victual, and officer, I shall really pity those fifty or sixty men."

The problems of waging war, about which most knew little, intruded even when momentous questions were about to be debated. On the day Richard Henry Lee presented the resolution for independence, the congress first spent some time discussing the quality of "powder manufactured by Mr. O. Eve's mill." Earlier, the congress had resolved that in order that America should get to know itself better, every colony be urged to create "a society for the improvement of agriculture, arts, manufactures, and commerce, and to maintain a correspondence between such societies that the rich and numerous natural advantages of this country, for supporting its inhabitants, may not be neglected.''

There were few businessmen in this congress dominated by lawyers and politicians, but those few, like Roger Sherman of Connecticut and Lewis Morris of New York, spent much time searching out ways to supply the army with tent cloth, shoes and clothing, saltpeter, sulfur, and gunpowder, and in the process learned a great deal about America's resources. Francis Lewis of New York, a signer of the Declaration but otherwise unheralded, seldom sat on committees that dealt with policy, but when the congress wanted a practical problem solved it often turned to him. In September 1775 it chose Lewis ''to purchase £5,000 worth [of] coarse woolen goods for the use of the Continental Army.'' In October it was thought ''expedient that I should . . . purchase necessaries for the troops at Cambridge.'' In February 1776 ''Mr. Lewis engaged to procure shoes for part of the army,'' a delegate reported. ''He has had a parcel made in Jersey because cheaper than elsewhere.'' While others agonized over independence, he searched for a way ''to forward to General Washington at Cambridge the five tons of powder now at New Brunswick.''

The congress tried to organize the gathering of information by creating a committee to ''purchase and furnish supplies for the army.'' Concerned citizens like Israel Gilpin sent in suggestions. ''As Congress is now a-promoting every useful art and science as well as preserving our liberties and properties,'' he wrote in December 1775, ''I have thought it might be of use to employ some honest skillful men to make search for sulfur, coal, etc., as it might be of immediate use and service and if a reconciliation should take place these things would be of infinite use to our posterity.'' A steady flow of reports from committees of public safety, designed to supervise each colony's military affairs and supply the militia with necessities, expanded the congress's knowledge of what America had and what it lacked in resources.

The most precious and perhaps most unexpected asset that turned up was a broad pool of skilled labor. Many were employed in an extensive iron industry—the colonies in 1775 produced thirty thousand tons of crude iron, or about one-seventh of the world's output—many more in an even more extensive shipbuilding industry, and still others in small workshops scattered

throughout the North. The talents of the artisans often went un-
used in the war effort or misused when most needed. Many, for
the lack of raw materials, closed their shops, mills, and forges
with the outbreak of fighting; others out of enthusiasm for the
patriots' cause joined the army. Nathan Sellers, a paper maker,
marched off to war in the summer of 1776. Soon after, the congress voted to issue Continentals, paper money to finance the war,
but could find no paper on which to print it. Someone had the
sense to track down Sellers; once found, he was recalled to civilian
life, and there he remained throughout the war. Leddeus Dod of
New Jersey, a maker of watches and surveying instruments,
sought glory as a captain of artillery and fought in a couple of
skirmishes, then Washington had him detached from service to
head up an armory. Paul Revere, silversmith and engraver, also
yearned for glory, but was quickly ordered down from his horse
and told to use his mechanical skill to repair cannons on Castle
Island in Boston harbor. (In the process he invented "a new type
of gun-carriage," but his biographer Esther Forbes fails to spec-
ify what was new about it.) Still later he went to Philadelphia to
learn how to make gunpowder and to "obtain an exact plan" for
constructing a powder mill. Oswald Eve was told to give "Mr.
Revere such information as will enable him to conduct the business on his return home," but Eve's loyalty to the cause had
limits. He rushed his visitor through the mill, leaving him no
chance to study the machinery or to talk with the workmen. Nonetheless, Revere saw enough, and with a floor plan of the mill
purloined by the intrepid Samuel Adams, he soon raised a powder
mill outside Boston. In a later venture he learned to cast and bore
cannons. It would appear that within a year or two after the
Declaration, whenever those with mechanical skills came to the
attention of the new nation's political and military leaders, those
talents were harnessed to the war effort wherever they could be
used.

Gunsmiths, of which there was no shortage, naturally received
early attention from the Continental Congress. Silas Deane reported in August 1775 that "the gunsmith's business goes on
well" in Connecticut, and Benjamin Franklin answered that "we
make great progress in it here." Similar reports came from Mary-
land and New York. But gunsmiths regarded themselves as art-

ists, proud of the style and individuality of the weapons they produced. The length of the barrel, the size of the bore, the shape of the lock varied from one artist to another. This diversity would not do for the army, which wanted a standard gun that took a standard ball. A congressional committee headed by Franklin discussed this matter with Washington in Cambridge in October 1775 and came up with what was perhaps the first suggested technological innovation of the Revolution:

> Agreed that it be recommended to the several assemblies or conventions of the respective colonies to set and keep their several gunsmiths at work to manufacture good firelocks with bayonets, each firelock to be made with a good bridge lock, three-fourths of an inch in the bore and of a good substance at the breech, the barrel to be three feet eight inches in length and a bayonet of eighteen inches in the blade, with a steel ramrod the upper lock to be trumpet-mouth'd. The price to be fixed by the assembly or convention or Committee of Safety of each colony.

How much success the committee had in imposing its directions on gunsmiths remains unclear, but the idea of a standardized product had been aired.

John Oliver exaggerates when he writes that "demands made by the war spurred the colonists on to great technological achievement—achievement which otherwise might have been postponed or never have occurred." David Bushnell, who has been called "undoubtedly one of the greatest mechanical geniuses this country has ever produced," constructed a submarine that failed. He had no more luck with floating mines. David Rittenhouse tried to develop a rifled cannon and with Charles Willson Peale experimented with a telescopic sight for rifles. Success eluded them. Oliver Evans created an ineffectual card-making machine. Thomas Paine's plan to set British warships afire with incendiary arrows came to nothing. But these and other aborted experiments revealed a trait of American artisans that was to help shape the future: they were always willing to apply their talents to new fields and to explore new ways of doing old things.

If few lasting mechanical innovations came from the Revolution, seeds were planted to blossom in the postwar period. Contact with French engineers—"We are in great want of good engineers,'' Franklin pleaded in 1775, men "acquainted with field service, sieges, etc., and ... with fortifying seaports,'' and the French army supplied the need—and with British soldiers who chose to stay in America opened a window on European achievements in technology. Another window opened to artisans who served, however briefly, in the army: travel and contact with brethren beyond the borders of their own "country" exposed them to new tricks of the trade, which they brought home after the war. Finally, America learned during the war that for all its abundance, it lacked much. "We want coals,'' Franklin had begged in 1775, and for the want of it many forges remained cold during the war. Likewise, after the war, those who worked with iron had as before to use coal imported from Liverpool. America felt sure early in the war it could produce all the saltpeter it needed for gunpowder, but it never did: most of it came, as before, from the Dutch and the French. America felt sure it could manufacture enough guns to equip its army, but it never did: most of the weapons used at the battle of Saratoga came from France. An attempt to mass-produce cloth in Philadelphia failed, and the Continental Congress ended up getting what it needed for stockings, uniforms, and tents from France. America knew itself better at the end of the war, but it also knew how much had to be done to free itself from dependence on Europe. Of course it could be, it would be done. "Great scenes inspire great ideas,'' Thomas Paine told his countrymen. "The nature of America expands the mind, and it partakes of the greatness it contemplates.''

Congress had earlier latinized into *Novus ordo secolorum* the Spirit of '76 that Paine had rendered into eloquent English. The words were apt for the Great Seal of the United States but meaningless to men like Paine, with only a grammar-school background, and to most Americans, until translated into English— "A New Order for the Ages.'' Many plain people of the day may have wondered how new the order would be if their leaders had to proclaim it in a dead language. Congress, without fanfare, did its best to live up to the ideal. It abandoned the inane British currency system and adopted one based on the decimal system the rest of Europe used. It sought to avoid "entangling alliances''— Paine's phrase from *Common Sense,* not Jefferson's, as generally supposed—but reluctantly accepted one when France insisted. The Confederation's congress in its dying years solved two problems England had boggled. The Land Ordinance of 1785 offered an orderly way to distribute land west of the Appalachian Mountains, and the Northwest Ordinance of 1787 created a way to bring those lands from territorial or colonial status into the Union as states with all the rights and privileges of the original thriteen. Also in 1787 came the newly minted Constitution, which two years later led to the creation of the federal government under George Washington's leadership.

Many at the Constitutional Convention had wanted to encourage technological innovation with "public institutions, rewards, and immunities for the promotion of agriculture, commerce, trades, and manufactures," but the only item in the final document that encouraged such advances gave to the federal government the right, which previously belonged to the states, to issue patents. Congress, seeking a new order for the ages, in 1790 passed a law to make the patent process easier and cheaper than in Great Britain. Only the well-to-do there could afford a patent; the cost varied from 70 pounds in 1815 to more than 150 in 1829. An American patent during the same period cost approximately 30 dollars, or in British currency a bit over 6 pounds. But the American law ultimately failed. The power to judge an application fell to the Secretary of State, an already overworked gentleman. When Jefferson held that office, he studied every application with care. He interpreted the law so strictly—an invention must be both workable and totally original—that he issued only thirty-seven patents during his tenure, most of them of little consequence. In 1793 Congress amended the law to let patents be issued without the test of originality or even usefulness, and allowed the federal courts to judge the validity of an application. The new law had a deadly flaw—the challenge of infringement cost money, and the artisans who submitted applications for patents had little of it to spare. It took until 1836 to remedy the defects.

While those in government worked to create a new order for the ages, the artisans in workshops up and down the land—the dirty-fingernail people, as they have been called—replicated in their own spheres the politicians' innovations. In the 1780s Oliver Evans produced, as noted, a card-making machine that led eventually to one that worked efficiently; then he created an automated grist mill that in time opened the way to the mass production of flour; and finally he invented a high-pressure steam

engine early in the next century, which powered steamboats up
and down western waters. In 1793 Eli Whitney invented the cotton gin, an "engine" that was to transform the South. In 1795
Jacob Perkins invented a nail-making machine, "the greatest
technological advance at the turn of the century," Thomas Cochran
has said. Imported wrought iron or American-made nails cost
twenty-five cents a pound in Perkins's day; within a generation
machine-made nails sold for eight cents a pound. In 1798 John
Fitch put a flawed steamboat on the Delaware River; it traveled
more than a thousand miles before collapsing, and led the way to
Fulton's achievement in 1807. John Kouwenhoven perhaps exaggerates when he writes that at the turn of the century "the unity
of no other nation in history rested to a similar degree upon
technological foundations," but he counts among the first historians to equate the quiet achievements that had sprung from the
underground world of the artisans of '76 with the much-honored
politicians of the day.

No one then or now seems to doubt that such an amorphous
thing as a Spirit of '76 existed. In 1853 Josiah Quincy remarked
on the enduring effect of that spirit. "Among the causes which
gave the impetus to the great improvements which this nineteenth
century has been distinguished, the principal has been, in my
judgment, the American Revolution. The common mind of the
time was set free to think, particularly in the United States, where
the mind was not hampered by the prejudices and unwieldy habits of former ages." Bernard Bailyn has echoed that sentiment in
the twentieth century: "Nothing so clearly documents the trans-
forming effect of the Revolution as the elevation of spirit, the
sense of enterprise and experimentation, that suddenly emerged
with Independence and that may be found in every sphere of life
in the earliest years of the New Republic." Historians of post-
Revolutionary era politics see the period as one of counterrevolu-
tion. Bailyn rejects this point of view. "Far from the 1780s being
a conservative or 'counterrevolutionary' period that culminated
in a Thermidor at Philadelphia and far from that decade being
dominated by self-searching despair for the future of republican
hopes, those years witnessed a vast release of American energies
that swept forward into every corner of life."

3

The Founding Fathers

"I worry that the 'dirty-fingernail' people are being squeezed out by the marketing and financial people," a builder of commercial airliners said not long ago. "We are losing that cadre of people who care first about the airplanes and their quality." Obviously, he did not mean to denigrate those he labeled dirty-fingernail people. Still, it does have a built-in deprecatory tone, and therein lies a problem—what to call these founding fathers. Inventors? That will not do. An account of the nuts and bolts of the American past in the century after the Declaration of Independence cannot be simply a history of inventions, though they must have a large place in the story. Engineers? That sounds too imposing, also a bit pompous for these people, somewhat like calling a janitor a building superintendent or a garbage collector a sanitary engineer. Yet it is tempting to call these men engineers. David Billington equates engineering with technology and notes that "engineering schools are often called schools of technology without implying any difference in courses of study at all." But the dirty-fingernail people, at least in the past, had little in common with engineers. When they met someone like Benjamin Latrobe, a cultivated Englishman who became a celebrated architect of American public buildings, they called him an engineer, meaning, as Latrobe put it, someone "considered only an overseer of men who dig, ... one that watches others who hew stone and wood." He got along much better with members of the elite like Thomas Jefferson than with the dirty-fingernail people, most of whom were literate but few of whom had been educated much beyond the three *R*s. The Sellers family produced an unbroken line of such people, dating back to the seventeenth century, but not until 1873 did one of them attend college. "We go forward without facts, and we learn the facts as we go along," Henry Ford once said, seeking to explain the difference between himself and the engineer. "There wouldn't be any fun if we didn't try things people said we can't do."

In the past the dirty-fingernail people generally referred to themselves as craftsmen or, more often, artisans, and for the moment let us consider them as such. Artisans, whether builders of airplanes today or furniture makers of an earlier century, whether trained in modern America or in eighteenth-century England, have through history, regardless of the times in which they live, the places they work, the tools they use, or the materials they handle, shared universal traits.

First, much as it would have appalled Sidney George Fisher to hear it said, they are thinking people. Their work, as Anthony Wallace puts it, "is in large part intellectual work." In the 1820s Nathan Sellers suggested to a Quaker colleague a way to make a particular operation more automatic. "Nathan," came the reply, "I have no doubt what thee proposes would act just as thee suggests; but when I hire a workman I hire his brains as well as his hands." In the 1970s Robert Pirsig said much the same thing: "The craftsman isn't ever following a single line of instruction. He's making decisions as he goes along.... He isn't following any set of written instructions because the nature of the material at hand determines his thoughts and motions, which simultaneously change the nature of the material at hand."

Second, they have an intimate, intuitive knowledge of the material with which they work. "My own eyes know because my own hands have felt," said a nineteenth-century craftsman who worked with wood, "but I cannot teach an outsider the difference between ash that is 'tough as whipcord,' and ash that is 'frow as a carrot,' or 'doaty,' or 'biscuity.'" A man who works with metal today speaks of the "mechanic's feel" for his material, "which is very obvious to those who know what it is, but hard to describe to those who don't; and when you see someone working on a machine who doesn't have it, you tend to suffer with the machine."

Nathan Sellers. Patriarch of a line of Philadelphia mechanicians, who was withdrawn from military service during the American Revolution to produce molds used to print currency issued by the Continental Congress. The portrait is by Charles Willson Peale. *(New York Public Library)*

Finally, craftsmen are artists—here again Fisher would scoff at the thought—and judge themselves as such. No one among them thought it pretentious to call Isaac Sanford the "chief artist of Middletown Manufactory." Pirsig reminds us that the root of "technology" comes from the Greek word for art. "The ancient Greeks never separated art from manufacture in their minds, and so never developed separate words for them. . . . a real understanding of what technology is—not an exploitation of nature, but a fusion of nature and the human spirit into a new kind of creation that transcends both." The word "art" appears often in the few writings the dirty-fingernail people have left behind. "One man's arms learnt (in the artist's way) to recognize what mattered," George Sturt remarks about the work that went on in his wheelwright shop in England :

Science? Our two-foot rules took us no nearer to exactness than a sixteenth of an inch. . . . So the work was more of art—a very fascinating art—than a science; and in this art, as I say, brain had its share. A good wheelwright knew by art but not by reasoning the proportion to keep between spokes and felloes; and so too a good smith knew how tight a two-and-half inch tyre should be made for a five-foot wheel and how tight for a four-foot, and so on. He felt in his bones. It was a perception with him. But there was no science in it; no reasoning. Every detail stood by itself, and had to be learnt either by trial and error or by tradition.

Walter Chrysler, a onetime mechanician turned industrialist, once said, ''There is in manufacturing a creative job that only poets are supposed to know. Someday I'd like to show a poet how it feels to design and build a railroad locomotive.'' As artists, the mechanicians always carried a pencil to illustrate their ideas when words failed. The old-timers, untrained as draftsmen, often used a blackboard for their canvas. Coleman Sellers had two large ones in the shops in which he built fire engines. ''On one of them was half of the set of levers drawn to full size for the large engines and on the other were the levers for the village engines which the blacksmiths worked to for the curves.'' They also thought like artists—visually. ''Father was very ready with the pencil and was one of the best offhand sketchers I ever knew, and he made good use of me in making the full size drawings,'' George Sellers recalled. ''In making these drawings part of the time I was obliged to lie on my belly and use my arm as radii for the curves with father standing by directing the changes of the trial marks I made.'' As a sculptor first makes a model before undertaking a project, so John Fitch's steamboat, Oliver Evans's steam-powered dredge, Patrick Lyon's fire engines, and Matthias Baldwin's first locomotive emerged from models built to guide their eyes as they proceeded. This custom continued into the twentieth century. ''Mr. Ford first sketched out on the blackboard his idea of the design he wanted,'' an associate recalled; then Charles Sorensen transformed the sketch into wood. ''Henry liked this. Never very happy with blueprints, he found it so much easier to work three-dimensionally, and he hired Sorensen—at $5 a day—to bring his pattern-making skill to the Ford Motor Company.''

These founding fathers, then, are artists. They are also crafts-men, but with a difference. Craftsmen, unlike most of the dirty-fingernail people who will figure in this story, are often hidebound, governed by handed-down traditions and determined to resist innovations. Also, the popular mind tends to connect craftsmen with simple tools—the hammer and saw, knife and chisel—seldom with machines. A more precise word is needed for our cast of characters. There is one that comes from the eighteenth century: mechanician, a person "who practices or is skilled in the mechanical art." Mechanicians are artisans (or artists) skilled in the operation, repair, and creation of machinery. They are a spe-cial breed that cultivated men like Sidney George Fisher never came to know or wished to know, yet they were the gentlemen who gave him his gas lighting, indoor plumbing, the "hoisting ma-chine," and a host of other objects that made his life increasingly more pleasant.

In the 1930s Hans Zinsser explained why he had become an epidemiologist. The field, he said, "remains one of the few sport-ing propositions left for individuals who feel the need of a certain amount of excitement. Infectious disease is one of the few genuine adventures left in the world. The dragons are all dead...." Any of the founding fathers could smile at the snobbish implications of the scientist's remark and yet agree that it expressed exactly their own feelings about their specialty. They worked constantly on the frontiers of knowledge, but, as Ford put it, "without facts." They were members of a small, elite international frater-nity that numbered at most in the hundreds, and they all knew one another by name and reputation. They shared a common love of tools and machines, which, along with their pencils, let them transcend cultural barriers. Those barriers, the Americans learned, loomed larger than they ever dreamed before they trav-eled to England.

The American mechanician, unlike his English counterpart, seemed to be a born peripatetic, constantly on the prowl for "sporting propositions." Sidney George Fisher thought himself a cosmopolite; compared to American mechanicians he was a pro-vincial snob. Isaac Markham, a backcountry boy reared in Mid-dlebury, Vermont, began as an apprentice to a local mechanician, moved on to the mills at Waltham, Massachusetts, to Paterson,

New Jersey, and finally back to Middlebury, where he became superintendent of a cotton mill. Joseph Saxton completed his apprenticeship in Philadelphia, then, "to enlarge his knowledge," went to England and "became acquainted with some of the outstanding engineers and scientists of his day," among them Michael Faraday. The extraordinary Jacob Perkins came down from Massachusetts to Philadelphia and there kept the mechanical world "in a feverish state of excitement. . . . It was never what he had done but what he was doing. . . . his schemes set many levelheaded men to thinking in the right direction." He, too, eventually went to England and stirred dull men alive. Thomas Edison roamed the country until finally, at the age of thirty-one and already known as "the wizard," he built himself a laboratory and settled down at Menlo Park. Henry Leland moved from shop to shop after he left the New England farm—from the armory at Srpingfield to Colt's factory in Hartford to Brown and Sharpe's plant in Providence ; he ended as "the grand old man" of Detroit, where he designed and built the engines for the original Cadillac.

An extraordinary number of these mechanicians—John Fitch, Joseph Saxton, Matthias Baldwin, Henry Ford, Joseph Brown, to name only a few—had in their youth been clockmakers or repairers of clocks. They carried over into whatever area they eventually made their specialty "a skill for making just about any iron piece," David Landes has said. "Indeed this versatility is a salient characteristic of the profession : anyone who can make so complicated and finely adjusted mechanism as a timekeeper can generally make any part thereof or anything else that uses the same materials." The experienced English mechanician, by contrast, tended to focus on specific, detailed work. Once trained to shape a rifle barrel or whatever, he rarely exceeded the bounds of his expertise. His tools—awkward-looking to Americans—were shaped to suit himself, no one else. The American mechanician, forced to wander in order to survive, had to accept the tools and patterns of work found in new shops. This necessity helped to make him a generalist—someone not only adaptable to new tools, new ways of doing things, but able to swing easily from one assignment to another. As the American doctor, unlike his English counterpart, diversified into amputations and the concoctions of medicines, so, too, did the American mechanician. Coleman Sellers's shop in Philadelphia at one time or another made paper molds, firehoses, fire engines, and carding machines, among other items. The men who came to work for him had to adapt to changing jobs that flowed into the shop.

The constant movement among mechanicians contributed to another distinctive trait within the American community—the lack of secrecy. While English entrepreneurs hugged their technology to the chest as if life and wealth depended on keeping it secret, Americans opened their doors to innovation. They felt that each man hired brought new, possibly useful, ideas to the shop, and no one tried to staunch the flow of information. There was an unwritten agreement that everyone had a right to know what the rest of the community was up to. Secrecy was impossible. One mechanician tended to marry into the family of another, which also helped to speed the flow of facts. The government armories encouraged open communication, and by 1840 all of the innovations established at Springfield, for example, had spread through the mechanical world of New England.

Lest too much be made of the uniqueness of the American mechanician, let a quotation from Condorcet, the philosopher-mathematician, whose experience was limited to France, remind us of the international character of the mechanician. He writes:

> Generally speaking, people have a very erroneous idea of the type of talent proper to the ideal mechanician. He is not a geometrician who, delving into the theory of movement and the categories of phenomena, formulates new mechanical principles or discovers undetected laws of nature.... The workshops, the machines themselves, show what has been achieved, but results depend on individual effort. To understand a machine it has to be divined. This is the reason why talent for mechanics is so rare, and can so easily go astray, and this is why it is hardly ever manifested without that boldness and the errors which, in the infancy of science, characterize genius.

If, as Fernand Braudel has said, technology, regardless of the national background, is ''a slow accretion of small changes over a long period,'' the mechanician has been a key agent in those small changes. Adam Eckfeldt and Joseph Saxton, both of Philadelphia and renowned among contemporaries but now largely forgotten, were exemplary mechanicians. Eckfeldt triumphed because "under his care many apparently slight improvements were gradually adopted that in the aggregate amounted to a great deal of economy of working." Saxton stood "neither behind nor in advance of his age, but in perfect harmony with it," an admirer said. "He neither pestered the world with premature projects destined to failure because the necessary contemporaneous conditions were not present, nor retarded the advance of improvement by advocating old errors under new forms."

4

The Making
of a Mechanician

Coleman Sellers owned a machine shop in Philadelphia passed on to him by his father and which he hoped to pass on to his sons. He began to entice them into the world of mechanicians before they were aware they might have a choice of other careers. George Escol, the second son, remembers that before he was ten his father built a boy's carpenter's bench in the garret alongside his own bench, where he often worked at night by candlelight. The boys had not yet reached their teens when their father bought a set of tools for them.

> I never can forget the glee of us boys when helping carry the tools up to the garret shop and to see the then empty tool chest hoisted up outside and taken in through the gable window. It was the advent of these tools that gave us our first lessons in the importance of system and order. With care and under father's watchfulness we were allowed to use some of these tools. It was an adage with father, ''That an indifferent workman might do tolerably good work with good tools, but that it took a skillful one to do so with bad tools.''

Gradually the father let the boys use his lathe, but it was too high and the boys too short ''for either of us to do the turning and at the same time tread the treadle.'' A stool solved that problem. One lad stood on it while the other worked the treadle. When they had mastered the basic mechanics of the machine, their father said the time had come when they should have one of their own built to their size, ''and long discussions came as to the ways and means of accomplishing this.'' Construction of the lathe led to visits to other workshops. Isaiah Lukens finished the lathe heads and rest carriers; Oliver Evans's shop cast and fitted up the crank wheel.

The fraternity of mechanicians joined with Coleman Sellers to draw George Escol into their world. A peripatetic German coppersmith in Sellers's shop taught as much of his art as the boy could absorb. (Coleman Sellers had been reluctant to hire a man with ''wandering habits,'' but the German said ''the more move the more learn and the better work do,'' and Sellers was forever glad he took him on.) Franklin Peale, a son of the artist Charles Willson Peale, was himself an artist with the cold chisel, and he took time while working in Sellers's shop to give ''many good lessons in handling and use of tools.'' Adam Eckfeldt, chief coiner of the United States Mint, had his own way of teaching the lad. One day he spotted him standing on tiptoe peering through the barred window of the coining room.

Seeing me peering over the bar, he took me by the arms and lifted me over it. Setting me down by the coining press he asked me if I did not want to make a cent. . . . He put a blank planchet into my hand, showed me how to drop it in, and where to place my hand to catch it as it came out; the lever and weights were swung, and I caught the penny as we boys called cents, but I at once dropped it. Mr. Eckfeldt laughed and asked me why I dropped it?

Because it was hot and I feared it would burn me. He picked it up and handed it to me, then certainly not hot enough to burn; he asked if it was not cold when he gave it to me to drop into the press; he told me to look and see there was no fire, and feel the press that it was cold; he then told me I must keep the cent until I learned what made it hot; then I might, if I liked, spend it for candy.

When I showed the bright new cent to my father whom I found in his workshop, and asked him to tell me what made it hot, he said he would show me; he handed me a common sulphur-

tipped match, then took up a small rod of copper, told me to feel
that it was cold, held its end on an anvil, and struck it a few
quick sharp blows with a hammer, then applied it to the match
which I held in my hand, which to my amazement was at once
lighted; he said, now you have something to think about and
may be able to understand when you are older; it was an object
lesson that led to many a train of thought.

The elder Sellers's workshop employed from ten to fifteen men.
The machine shop held lathes, a workbench with a grindstone and
a vise or two mounted on it, a tool chest filled with hammers, files,
chisels, and wrenches. There were four large hearths in the blacksmith's shop with a small one for dressing tools. Youngsters assisted the smiths and pumped the bellows that kept the hearths
blazing. In the pattern room carpenters shaped wooden forms for
casting. "It is curious how ... the first fire engines were gotten
together," George Escol Sellers recalls. Much of the work was
subcontracted to other shops. John Wiltbank cast the cylinders,
Israel Morris did the copper work, Oliver Evans forged and fitted
the pumping levers. "The ash wood handles were turned by Han-
sell, the wheels and running gear at a carriage shop at Heston-
ville.... I must not forget the painting, which was done by
Woodside, the sign painter." If the quality of work failed to
satisfy Sellers, the subcontractors heard from him.

At the back of the shop was a waste scrap room and space for
John Brandt, whom Sellers had put on the payroll while he
worked on the card-making machine he had invented, which
Sellers was to build and sell when it was perfected. (Wallace gives
as a partial explanation for the advance of British technology in
progressive firms their "enduring physical facilities, with enough
redundancy so that a niche for experimentation was always available, safe from the demands of day-to-day operations.) At the
front was Sellers's office, where friends assembled at the end of
the workday to talk shop—mechanicians like Patrick Lyon, Matthias Baldwin, Isaiah Lukens, and Rufus Tyler; fellow travelers
like Dr. Thomas Jones, erstwhile superintendent of the patent
office and later editor of the Franklin Institute's journal; Professor Robert Patterson, a mathematician and also director of the
mint, and Dr. James Mease, whose account of Philadelphia in

1811 included a survey of its machine shops. If the space resembled the one in Pat Lyon's shop, as it probably did, it was "a little cubby-hole of an office, warmed by an old wood burning tin plate stove, around which he and his men would congregate of evenings to crack jokes, smoke their pipes and drink their grog."

George Sellers's formal education ended when he was fifteen and went to work for his father. He did odd jobs about the workshop, running errands—"With all this scattered work you can form some judgment of the amount of leg wear to a boy old enough to run errands"—paying subcontractors who brought in finished work, weighing the wire for cards the women and children who trooped into the shop would assemble. His father enrolled him in a course on mechanical drawing; otherwise the old-fashioned introduction into the mechanician's world continued uninterrupted in the midst of pedestrian tasks. Father and father's friends and colleagues educated him by example, by practice, and by exposure to new ways of doing old things. The lessons they taught the boy Henry Ford expressed succinctly more than a century later when he advised an apprentice, "Find out what was wrong, try to understand why it had gone wrong, and then break down the corrective process into modest steps."

What sort of man emerged, still emerges, from this training? Someone, as Pirsig puts it, who has a feeling for the quality of his work, especially when stuck on the solution to a problem. "It is this understanding of Quality as revealed by stuckness which so often makes self-taught mechanics so superior to institute-trained men who have learned how to handle everything except a new situation." Pirsig goes on to describe the sort of man Coleman Sellers sought to make of his son:

> If you have a high evaluation of yourself then your ability to recognize new facts is weakened. Your ego isolates you from Quality reality. When the facts show that you've just goofed, you're not likely to admit it.... On any mechanical repair job ego comes in for rough treatment. You're always being fooled, you're always making mistakes, and a mechanic who has a big ego to defend is at a terrific disadvantage.... mechanics tend to be rather modest and quiet. There are exceptions, but generally, if they're not quiet and modest at first, the work seems to make

them that way. And skeptical. Attentive but skeptical. But not
egoistic. There's no way to bullshit your way into looking good
on a mechanical repair job, except with someone who doesn't
know what you're doing.

But let Coleman Sellers have the last word on the making of a
mechanician, which could serve as an epitaph for all who have
sought to attract an upcoming generation into their profession:
"He used to tell us never say 'can't' but to try and keep on trying
until we could."

5

Fellow Travelers

At least 80 percent, probably more, of the technological innovations down to the Civil War came from the workbenches of mechanicians. But these gentlemen had allies who seldom dirtied their hands, yet admired, listened to, accepted, and promoted the mechanicians and what they produced from their shops. Inventors like Robert Fulton, an artist by training, leaned on them to bring a dream into reality. There were capitalists with cash to spare, like Robert Livingston of New York, who backed Fulton; the Browns of Providence, who subsidized Slater's mill for cotton yarn; Boston businessmen who invested in Lowell's mill at Waltham. Even government officials, state and federal, sometimes advanced taxpayers' money to finance the experiments of mechanicians in whom they had faith.

Among the closest allies were small-time entrepreneurs. Many mechanicians—Oliver Evans, Coleman Sellers, Simeon North, among others—doubled as entrepreneurs, but their culture, Anthony Wallace notes, differed fundamentally from the typical entrepreneur or manufacturer: "The machinist thought with his hands and eyes and when he wished to learn to communicate he made a drawing or a model; the manufacturer and manager thought with his larynx, as it were, and when he wished to learn or communicate, did so with words, in conversation or in writing. The machinists had dirty hands from working with tools; the manager had cramped hands from writing." Often the entrepreneur, an outsider who knew little about machinery, looked at the technical aspects of manufacturing through a fog. This drove him to depend on the mechanicians. Samuel Vaughan Merrick, a Philadelphia merchant, through happenstance inherited a bankrupt shop that built fire engines, a business about which he knew nothing. John Price Crozer, after losing the family farm and failing in other enterprises, bought a decrepit cotton mill along Brandywine Creek in 1821. Later he said, "It seems to me almost a miracle that I was not crushed at the very outset. It was difficult to get any one that understood the business." His machinery "was very inferior, and I an entire novice." He began to succeed after hiring an experienced mechanician trained in England to manage the mill. Joshua Gilpin's paper mill along the same creek prospered mainly because he enticed Lawrence Greatrake, privy to the latest techniques for making paper by machinery, to emigrate from England to run the mills. These entrepreneurs were propelled into the mechanicians' world by a single urge—desire for wealth. They accepted innovations if they increased profits. They were usually ill-informed about the machines in their mills or shops and depended on mechanicians to choose those improvements which would help to increase profits. They were, in short, conservative gentlemen concerned mainly with the "bottom line," but also adventurers who had often failed in some earlier enterprise. They tended to buy up, on borrowed money, idle mills and shops and transform them into profitable enterprises.

Another group of fellow travelers, most of whom knew little firsthand about machinery, promoted the mechanicians' work as unpaid publicity agents. George Washington headed this diverse group. He visited workshops and mills on his tours of the nation and praised their owners for helping to free America from dependence on Europe; he encouraged mechanicians like James Rumsey, who seemed to be on the verge of producing a workable steamboat. Tench Coxe, assistant to Alexander Hamilton, served as the ideologue of technology. In a stream of essays, speeches, and pamphlets he made it seem almost a duty to God to promote it. In the words of Leo Marx,

> It was Coxe who first gathered . . . scattered impulses and ideas
> into a prophetic vision of machine technology as the fulcrum of
> national power. . . . he was intelligent enough not to conceive of

the American power as emerging from technology per se, but rather from the peculiar affinities between the machine and the New World setting in its entirety : geographical, political, social, and, in our sense of the word, cultural.

Charles Willson Peale, one of the few of the group who was well acquainted with mechanicians, gave over part of his museum as a "repository of machines." Thomas P. Jones, an English emigrant who, as noted, served briefly as commissioner of the Patent Office, promoted the mechanicians' work in public lectures and as editor of the Franklin Institute's journal. All these gentlemen, though politically divided, united in their despair of Americans' "debauched taste" for European goods. (This debauched taste extended all the way to George Washington, normally a "buy American" advocate; he wore a suit of homespun at his first inaugural, but when he wanted a watch "well executed in point of workmanship" he sent to Paris for it.) Encouragement of American industry and thereby of its mechanicians, Coxe shouted to a deaf audience, "will lead us, once more, into the paths of virtue, by restoring frugality and industry, those potent antidotes to the vice of mankind, and will give us real independence by rescuing us from the tyranny of foreign fashions, and the destructive torrent of luxury."

The various interests of these fellow travelers, each with his special tie to the world of mechanicians, combined in one man— Thomas Jefferson. At first glance Jefferson's attachment to technology seems frivolous. Mechanical gadgets fascinated him, and he remained a hopeless addict to them all his life. "You know I had a wagon which moved itself," he wrote to a friend in 1770. "Cannot we construct a boat then which shall row itself?" Some of the gadgets he created were practical, like the storm windows that warded off winter winds and the portable camp chair in the shape of "a walking stick . . . composed of three sticks which being spread out and covered with a piece of cloth made a tolerable seat." Others—a dumbwaiter, concealed in a side panel of the fireplace, which brought wine up from the cellar; French doors that swung open together with a light touch to the handle of one; living-room windows that became passageways to the outdoors when the lower sash was raised; a weathervane on the ceiling of the front porch—were more clever than practical. When he heard about the pedometer, he bought one to measure his walks, then an odometer for his carriage. He purchased as soon as he learned about it a polygraph, a machine that duplicated by mechanically linked pens whatever he wrote—a purchase, incidentally, for which posterity is grateful, because much of his surviving correspondence consists of copies made by the polygraph.

Jefferson's fascination with gadgets led directly to mechanical devices that cut labor in "a country where there is more to do than men to do it" and that were "applicable to our daily concerns." By the time of the War of 1812 his plantation had become a community of machines as well as people. His pride was a moldboard plow he had designed on mathematical principles. Friends praised it, but "how useful the invention actually proved to be is a matter of conjecture," according to Merrill Peterson. Another showpiece was a thresher probably imported from England. "This machine, the whole of which does not weigh two thousand pounds, is conveyed from one farm to another in a wagon, and threshes from one hundred and twenty to one hundred and fifty bushels a day." He had a drill plow that opened a furrow, sowed seed evenly, and covered it, and cost only nine dollars. Jefferson also had a nail-making machine, a sawmill, and an automated grist mill based on Oliver Evans's design. He hoped that before the War of 1812 ended he would be making all his own cloth:

> my household manufactures are just getting into operation on the scale of a carding machine costing $50 only, which may be worked by a girl of twelve years old, a spinning machine, which may be made for $10, carrying 6 spindles for cotton, and a loom, with a flying shuttle weaving its twenty yards a day. I need, 2,000 yards of linen, cotton, and woolen yearly to clothe my family, which this machinery, costing $150 only, and worked by two women and two girls, will more than furnish.

It comes as a surprise to learn that Jefferson, champion of the yeoman farmer and rural values, also championed machinery, mechanicians, and technological innovation. The explanation is simple. The War of 1812 convinced him that America "had no

choice but to manufacture for herself,'' but she must do it, in
John Kasson's words, in a way that ''would not threaten, but
rather enhance, the independence and virtue of American society.'' The steam engine lay at the center of Jefferson's vision.
Watt and Boulton's ''large and expensive machinery,'' which ran
the ''dark, satanic'' English cotton mills that had appalled
Jefferson, must be reduced in size ''to the compass of a private
family,'' to serve daily needs, ''the small and numerous calls of
life.'' Oliver Evans's high-pressure engine gave America just the
machine it needed in Jefferson's eyes. It was small and powerful
enough to propel a steamboat upstream yet leave plenty of room
for cargo and passengers. It was also portable and adaptable to a
variety of other needs. Jefferson saw that someday it might be
used to pump water from a well to a rooftop and there, stored in
tanks, supply running water for a family or serve as a ''resource
against fire.'' In time it might be adapted ''to wash the linen,
knead the bread, beat hominy, churn the butter, turn the spit, and
do all other household offices which require only a regular mechanical action.'' In Jefferson's eyes, Leo Marx has said, ''the
machine is a token of that liberation of the human spirit to be
realized by the young American republic. . . . he assumes it will
blend harmoniously into the open countryside of his native land.
He envisages it . . . helping to transform a wilderness.'' Neverthe-
less, Marx adds, ''it is the intensity of his belief in the land, as a
locus of both economic and moral value, which prevents him from
seeing what the machine portends for America.''

II

The Appearance
of Things Unfamiliar

When Rip Van Winkle returned home sometime in the 1780s after a twenty-year slumber in the Catskill Mountains, he found that the portrait of George III that hung outside the local inn had been altered. "The red coat was changed for one of blue and buff, a sword was held in the hand instead of a scepter, the head was decorated with a cocked hat, and underneath was printed in large letters, GENERAL WASHINGTON. The substitution of General George for King George marked the sharpest change Rip had to absorb. Friends had aged or died but otherwise the village remained much as it had been.

The sameness of life from one generation to another began to change about the time General Washington became President Washington. If Rip had slept through the next twenty years, he would have found on awakening his old familiar world visibly altered. Women had put away their spinning wheels and were buying thread and yarn made by machines. People and cargoes were moving up and down streams in noisy steamboats rather than silent sailboats. In some parts of the country mills were turning grain into flour without the intervention of human hands. Clocks, once made of brass by skilled craftsmen and sold mainly to the rich, were now being made of wood, mass-produced, and priced cheap enough for a backwoods farmer to afford. Rip could still see many familiar things—houses, boats, and wagons built as they had been—but the host of unfamiliar, even mysterious things would leave him flabbergasted, feeling like a stranger in his own land.

6

The Brain Drain Begins

No two mechanicians could have differed more than Oliver Evans, with his large leaps into the unknown, and Samuel Slater, something of a plodder, who took few risks and made few innovations as he transferred to America the British technology familiar from his youth. But there is room in the world for both the tortoise and the hare. Slater in his own way left a mark on American technology as deep as Evans's.

Slater, reared in England, was fourteen when Jedediah Strutt, a friend and neighbor of his father, took him on as an apprentice in his factory, which housed yarn-making machinery designed by Richard Arkwright. Strutt saw he had acquired a bright, literate, alert hand and soon, despite the boy's age, made him supervisor of the factory, a post that gave the young man a knowledge of the entire operation. Slater's apprenticeship ended in 1789. Strutt had retired, leaving others to run the factory but keeping the profits to himself. Slater saw that at the age of twenty-one he had gone about as far as a workingman in England could go. He emigrated to America. He took a stopgap job in New York in a textile mill but saw no chance of advancement there. One day he heard that Moses Brown, a wealthy Rhode Island Quaker merchant, was trying to produce cotton thread with Arkwright's machinery but having little luck. "We are destitute of a person acquainted with water frame spinning"—that is, a spinning machine powered by water—Brown wrote in answer to a letter from Slater. "If thy present situation does not come up to what thou wishest, ... come and work ours and have the *credit* as well as the advantage of perfecting the first watermill in America.''

Slater came to Pawtucket, a hamlet outside Providence, and looked over Brown's mill. The machines ''are good for nothing in their present condition,'' he said, ''nor can they be made to answer.'' The mill should be rebuilt from the ground up. He must have been an impressive as well as an agreeable young man, for William Almy and Smith Brown agreed to finance the project, with Moses Brown as a silent partner, and promised Slater 50 percent of the profits once it was operating. Despite Slater's inti-

Slater's spinning machine adapted the spinning wheel to mass production. It ''made forty-eight lengths of yarn simultaneously and did so without any human skill having to be applied,'' in Hindle and Lubar's words. ''The operator's job was to keep the machine running, keep it supplied with roving [carded cotton slivers] and fix it whenever a length of yarn broke or any other breakdown occurred.'' *(Smithsonian Institution)*

mate knowledge of Arkwright's machinery, he would have gone nowhere without the aid of local craftsmen: Sylvanus Brown (no relation to the eminent Brown clan of Rhode Island), a wheelwright who made the wooden machinery parts from Slater's drawings; Oziel Wilkinson, a blacksmith, whose shop turned out anchors, shovels, scythes, and other tools, who forged the parts that called for iron; and Pliny Earle, who made the castings for the carding machine. So much has been made of Slater's achievement that many forget what Brooke Hindle notes: "A country less developed technologically could not have carried through Slater's designs as they did." With their help and that of other locals like the leather workers who made the belts that turned the machines, Slater had the mill going on 20 December 1790. He began with four workers and in a month, with everything ticking along nicely, added five more. In less than a year and a half he was producing more tight, sturdy, uniform yarn than weavers in and around Pawtucket could use.

Soon after the wheels in Slater's first mill had begun to turn, Moses Brown and Oziel Wilkinson had bought land a couple hundred yards upstream for a new, enlarged "factory house." A dam, "one of the largest dams, if not the largest, yet built in America," was soon near completion. (Incidentally, two of the numerous local workmen were blacks, Prince Kennedy and Mingo.) It consisted of "fifty ox-cart loads of hewed timber, four thousand feet of two-inch plank, and one thousand weight of wrought iron." Three craftsmen from the hamlet soon wrecked it, even though the flow of water down to their own waterwheels remained the same. For them and others the dam and the mill to come "symbolically threaten" a deep-rooted way of life in the country hamlet. Besides, there was a prejudice against the Englishman Slater, a prejudice that according to a contemporary "lasted some time and attached to everything pertaining to cotton manufacture." (Slater's marriage to one of Wilkinson's daughters may have helped to alleviate the resentment against him.)

Within a few weeks the dam had been rebuilt and the two-and-a-half-story "factory house" began to rise along the river. It stands today, open to visitors, and known as the "Old Slater Mill." It was, and is, "a modest, simple, and unassuming affair," mainly because that was all the local workmen were capable of constructing. "What took place inside the factory was different than anything that had ever taken place in Pawtucket," Gary Kulik remarks in an excellent essay, "but visually the factory blended easily into the village landscape. There was nothing jarring about its visual presence, nothing in the image it projected likely to provoke further animosity. Indeed, given its style and proportions, it resembled nothing so much as an unadorned, eighteenth-century New England meeting house—an image of sober rectitude."

Once the new or "old mill" was working, Slater and his backers moved ahead cautiously. In 1793 they spent only $1,565, half for cotton, the rest for wages and other overhead expenses. The small operation stayed small for a decade. In 1803 the net profit was $18,000, half of which went to Slater. By this time he felt secure enough to branch out. He brought over a brother from England and together they built a second mill. By 1815, through partnerships like his original one with Almy and Brown, he shared in the profits of a score of mills up and down the Pawtucket River. In the process Slater transformed the countryside. The presence of a new mill rippled outward to touch every household in the area. "A new animation has been awakened," Timothy Dwight, a contemporary, said:

Common laborers, diggers of canals, lumber merchants, dealers in hardware, brass and iron founders, burners of lime, carpenters, masons, curriers, wagoners, sellers of wood, and blacksmiths are all employed in greater or less degrees by the erection of a cotton manufactory. To these are to be added the superintendents, clerks, overseers, agents at home and abroad, dyers, and that numerous class of men, women, and children who are immediately engaged in manufacturing the yarn. What is perhaps of still more consequence to the general prosperity, the weaving is all done in private families; and being spread throughout a circumference of sixty miles to the northeast and west of Providence, engrosses a number of which it would be difficult to estimate. The agricultural interest is estimated by the rise of land, the rise of produce, and a nearer and readier market. For example, a piece of land on a millstream fifteen miles from Providence was sold lately for fifteen hundred dollars an acre, which fifteen years ago would scarcely have been sold for one hundred. A manufactory of fifteen hundred spindles will soon accumulate a population sufficient to form a village.

The effect of Slater's mills extended far beyond Rhode Island. Through his own example and that of the Englishmen he recruited to run his mills, "he acted as the node from which Arkwright technology spread through New England and northern New York and even to Philadelphia."

Slater knew he had brought a revolutionary technology to America, but he did all he could to disturb as little as need be the cultural setting he had stumbled into and stirred up. Marriage to Oziel Wilkinson's daughter Hannah cemented him into the New England way of life. Originally, he tried to transplant the English apprenticeship system and hire youngsters to run his machines. When he saw Americans did not want their children detached from their families, he created new hamlets and recruited families to the spot. The women and children worked in the mills while the fathers tilled farmland leased to them by Slater. Wages from the mill were paid to the fathers, which left them with their authority to dominate the family intact. Only when the youngsters matured and became intrigued with machinery did their parents find it hard to keep them down on the farm.

Slater's influence on American technology cannot be exaggerated, but his ingenuity as a mechanician has been. Few innovations in the making of thread and yarn by machinery came from him. He was fundamentally "a close copier of well-tried British techniques." His success in carrying British achievements across the ocean has obscured the innovations of his remarkable father-in-law and his son, David Wilkinson. Both men worked amicably with Slater, whose English background helped to expand the realm of the possible for them. David proved to be an especially ingenious mechanician. His shop in Pawtucket, Theodore Z. Penn has said, "may well have been the first independent machine shop for building textile machinery in the United States. If not the earliest, it was one of the most important for supplying the needs of the growing textile industry." With his father he designed a grinding machine to make spindles for Slater's "old mill," but like Oliver Evans, he had talent and imagination that drove him to take larger leaps. In 1794 he created a power-driven lathe to cut screws for industrial presses. In 1800, at the age of twenty-nine he founded his own machine company and began to build machinery for canal locks, drawbridges, and steam engines of substantial size. Of his achievements Robert Woodbury has said, "In Wilkinson's shop was trained a whole generation of machinists and machine tool builders who laid the foundation of the use of machine tools in the United States. We may credit him as founder of the American machine tool industry and a contributor of the first rank of the industrial lathe."

Samuel Slater. *(New York Public Library)*

Slater was a remarkable man, intelligent, flexible, and generous with his knowledge. He adapted to America's unique environment quickly and deserves the attention historians have given him. He and a few others exemplified the best in the early brain drain. From France came E. I. du Pont, who built a powder mill in 1802 along the Brandywine based on technology acquired at home. From Germany came Henry Voight, Fitch's indispensable mechanician. Then, there were those from England, among them Lawrence Greatrake, an expert in paper-making machinery; George and Isaac Hodgson, mechanicians who escaped British surveillance by wrapping their tools in bundles labeled "fruit trees"; Benjamin Latrobe, architect and engineer; Charles Stoudinger, a former employee of Boulton and Watt who helped to guide the development of Nicholas Roosevelt's steam engines; and William Weston, who gave freely of his expertise in canal building.

These gentlemen were the elite who brought brains and wide experience to America. The typical English mechanician contributed little to the advance of the new nation's technology. Du Pont commented bitterly, "Englishmen knows every thing better than every body." A New England entrepreneur said, "This establishment suffered much in the outset, in being put to much expense by English workmen who pretended to much more knowledge in the business than they really possessed. At present only two are employed, and Americans, as apprentices, etc., are getting the art very fast." He spoke in 1806. More than a century later Henry Leland, one of America's greatest mechanicians, recalled his experience as an apprentice in another New England shop where most of the workmen were English: "What can you expect of a Yankee?" they hooted as he went by.

The typical English mechanician often proved of less worth than any expertise he had to pass along. Many were addicted to alcohol. Others lied about their skills and soon proved useless. In the textile industry a large majority, perhaps up to 75 percent, only "possessed obsolete skills, predominantly hand loom weaving," David Jeremy remarks. "Relatively small numbers of workers with the new industrial skills—dozens rather than hundreds—reached the United States from Britain each year in the early nineteenth century." But entrepreneurs had to be wary even with those knowledgeable about machinery. Many were secretive and disclosed only enough of their skills to make themselves think they were indispensable to their employers. There were further drawbacks to hiring English workmen. Too often they came with limited experience and knew only how to manage a particular aspect of a complex operation. Worse, regional variations in British technology could create chaos when immigrants from diverse backgrounds rubbed shoulders in an American shop or mill. Said one commentator on American technology in the 1820s,

> I cannot conceive a more uncomfortable situation than, for a manager who is not perfect in the business, to be surrounded by a mixture of Irish, Yorkshire, and west of England workmen. Whatever advice he might receive from the one party would be condemned by the others. If any description of machinery were recommended by the one, the other would be sure to suggest something different as being better; their opinions . . . would be all at variance with each other.

The contributions of the brain drain, so far as the typical European workman goes, has been exaggerated. A few notable emigrants, among whom Slater leads the list, gave much to American technology, but most of those who came gave little.

7

A Practical Visionary

Oliver Evans called himself "a plain mechanical-minded man," but he was more than that and knew it. He had a first-rate mind that led down paths more conservative mechanicians such as Eckfeldt and Saxton avoided. The slow accretion of small changes over a long period did not attract him. He was a practical man who made a comfortable living from his Philadelphia machine shop, the Mars Works, but he also pestered the world of his day with visionary projects. In 1805, for instance, he remarked that "if an open glass be filled with ether and set in water under vacuo, the ether will boil rapidly and rob the water of its latent heat till it freezes." This idea led to the thought that water frozen by volatized ether could provide Americans with an endless supply of ice. "The first precise scientific vision of how cold might be mechanically produced and utilized" came from Evans, Siegfried Giedion has said. "He is, in conception at least, the father of modern cold-making."

Evans was reared on a farm in Delaware. During a harvest season a slice from his scythe made him useless in the fields, but it left him free to indulge a fascination with machinery. He might have been expected to create a machine that made reaping grain less dangerous and more efficient; instead, he set out in 1777 to build a card-making machine, when he was twenty-two. He aimed to combine into a single mechanical operation four currently performed by hand—cutting wire into teeth, bending the teeth to a desired angle, punching holes in leather cards, then setting the teeth into the holes. If the machine worked as conceived, hundreds of women and children in Philadelphia alone who assembled cards would have to look elsewhere for part-time jobs, but meanwhile Evans would have made enough money from the invention to escape the drudgery of farm life.

His father called him "cracked" as he watched the weird apparatus evolve. The whole family "united argument with ridicule to dissuade him from his visionary schemes." But the machine got built and, as Evans tells it, "succeeded so well that they all changed their language, and nothing could surpass the ingenuity of Oliver." He does not add that the machine failed, as remarked earlier, to supplant hand assembly, for the pierced holes "were too large to set the teeth firmly in the leather cards."

Evans's next creation was another visionary scheme—an automated grist mill, which he conceived about 1782 when the American Revolution was winding down. Any farm boy who had hauled grain to the local mill knew the routine that turned it into flour. A man carried the sacked grain up to the top of the mill. There he poured it into a trough that let it drift at a regulated rate down to the millstones below. After being ground, the damp meal spilled into another trough where another man shoveled it into hoist tubs that were hauled upward to the drying room floor. There still another man shoveled it from the tubs and spread it out by rake to cool and dry. Once dry the flour was bolted—the chaff sifted out—then packed into barrels, ready for market. It took from four to eight men, not counting the miller, to complete this process.

Evans's automated mill reduced the crew to two. The first man dumped the incoming sacks into a spout from which, after the flour was weighed, an endless leather bucket belt carried the grain to the top floor. From there other belts and "endless screws" moved it along—downward to the millstones, upward to the "hopper boy," where revolving rakes spread the meal to dry, downward again to the "bolting chest" and from there into a spout that filled flour barrels, while a second man nailed on the lids. Jacques Pierre Brissot de Warville, soon to die in the French Revolution, visited an Evans mill in 1788. He found it "admirably clean and well organized" and admired the use of "gears and mechanical devices, so as to save labor in operations such as hoisting the wheat, cleaning it, raising the flour to the place where it is to be spread, dropping it into the room where it is put into barrels, etc.''

Evans had adapted his creation to American conditions without changing the century-old pattern of milling. Water still powered the mill, but the wheel that previously only turned the millstones now moved all his machinery. The gears, axles, and screws were made of wood, an inexpensive item compared to the iron equipment favored in English textile mills. Leather belts might wear out, but they were also cheap and easy to replace. Historians today credit Evans with originating the modern assembly line. It was he, Siegfried Giedion writes,

> who first incorporated the three basic types of conveyor, as still used today, into a continuous production line.... The ''endless belt'' (belt conveyor), the ''endless screw'' (screw conveyor), and the ''chain of buckets'' (bucket conveyor), which he used from the very start, constitute to the present day, the three types of conveyor systems. Later these three elements became exhaustively technified in their details, but in the method itself there was nothing to change.

No one, then or now, disputes Evans's ingenious achievement. It speeded up production from perhaps one hundred barrels of grain an hour to three hundred or more, while cutting labor costs at least in half, probably more. It turned out a standard and also a clean product. Evans had been shocked, so he said, by ''the great quantity of dirt constantly mixing with meal from the dirty feet of everyone who trampled it, trailing it over the whole mill.'' This problem, coupled with his eagerness to make machines do the dull, routine work so often forced on unskilled labor, probably instigated his design of the automated mill.

But for all its virtues, contemporaries regarded the mill as a visionary scheme, which indeed it was in the 1780s. Millers then worried little about labor costs or the shortage of help because milling took place after the harvest season when farmers and their sons, with time on their hands, were eager to hire themselves out for jobs that paid little but called for no skill. It was years before the savings in labor would pay for Evans's machinery. And what if a single leather belt broke during the milling process? All must come to a stop until someone repaired the dead conveyor. Millers from around the Delaware Valley came to study the creation, and after standing with awe as the rumbling, creaking machinery produced superfine flour without human intervention, they returned home still skeptical. Evans's brother traveled through the Middle States but could convince no miller to install the automated mill. But in 1816, after another long trip through wheat-growing country, he saw flour being made everywhere by Evans's machinery. ''I have walked through mills calling for the miller and found none and the whole process of grinding, elevating, cooling and bolting going on and no miller.'' By 1837 there were at least twelve hundred automated mills in the country. Few of the owners paid Evans any royalties. They pirated his ideas from a handbook, *The Young Mill-Wright and Miller's Guide,* which he had published in 1795 to publicize his invention. It became something of a bestseller and by the eve of the Civil War had gone through fifteen editions.

Also by the eve of the Civil War, thanks first to Evans's innovation and later to McCormick's reaper, ''flour milling was the leading American industry,'' Ruth Schwartz Cowan remarks; ''the value of its product was $249 million, more than twice the value of the product of the cotton industry ($107 million) and three times that of iron and steel ($73 million).'' Most of it was superfine white flour, which did not deteriorate because the germ and bran had been ground from it. Cowan suggests that this surge in the production of mass-produced white flour, which the health addict Sylvester Graham of Graham cracker fame deplored, affected the daily lives of families in a still largely rural America. It forced them to make, ''in one, not insignificant aspect of their lives, the crucial transition from being producers to being consumers, from being involved with the product (grain) at almost all stages of its preparation to encountering it only at the very last stage—and acquiring it only through trade. They would have, in short, begun the first stages of the industrialization of their household.''

A decade before installing the first automated mill, Evans had become obsessed by another dream—creating a wagon driven by steam. He conceived a high-pressure engine more compact and economical than the Boulton and Watt low-pressure engine. In 1786 he asked the Pennsylvania legislature to issue him monopoly rights to both his automated mill and his high-pressure engine. The legislators approved of the mill but treated the engine ''with contempt little short of insult.'' They could not see ''beyond their noses,'' Evans said later. ''They could see grain go into the mill and come out flour, but as to a wagon being moved by any other power than the slow-moving ox, the horse, or mule, or being dragged by man power, was beyond their comprehension.'' Then in 1805 he built a steam dredge to clean out the silt that had collected around the docks of Philadelphia harbor. James Thomas Flexner's description of the dredge's D-Day cannot be bettered:

> He set the dredge waddling like some prehistoric monster into the heart of Philadelphia. For several days the ponderously named *Orukter Amphibolos* (amphibious digger) inched thunderously round and round the waterworks in Center Square, while the impoverished inventor passed the hat. Then the monster dragged its vast bulk to the Schuylkill and wallowed into the water. Hooking up a temporary stern wheel and steering with an oar, Evans navigated the river for a few hours before he anchored his barge and set it to its plebeian task of pulling up mud in buckets.

Benjamin H. Latrobe, who had designed the waterworks that the dredge circled in its first public appearance, laughed at the monstrosity and called Evans ''a visionary seized with steam mania.'' Mania it was. Someday, said Evans over and over to anyone who would listen, men would be ''not only navigating our rivers, but crossing oceans and continents by steam power.'' He reflected his bitter reaction to the ridicule heaped upon him in the title of a brief book he published in 1805, *The Abortion of the Young Steam Engineer's Guide*. The dredge soon vanished from the scene, but Evans continued to tinker with his engine. Within a short time he had it driving lathes and other tools in his Mars Works, and later the engine or pirated models of it were sending steamboats up and down western rivers.

A year or so before he died, Evans rode out to the countryside to a grist mill where he was installing automated machinery. In the buggy were his friend Coleman Sellers and Sellers's ten-year-old son George. The youngster listened raptly as the old man reminisced :

> Mr. Evans had much to say on the difficulties inventive mechanics labored under for want of published records of what had preceded them, and for works of reference to help the beginner. In speaking of his own experience, he said that everything he had undertaken he had been obliged to start at the very foundation ; often going over ground that others had exhausted and abandoned, leaving no record. He considered the greatest difficulties beginners had to encounter was want of reliable knowledge of what had been done.

He thought a Mechanical Bureau ought to be created to "collect and publish all inventions, combined with reliable treatises on sound mechanical principles, as the greatest help to beginners." Mechanicians must support such a bureau, but also manufacture, merchants, and other fellow travelers.

Condorcet, writing years earlier from France, would have agreed with Evans—up to a point. "If a scholar poses himself a new problem he can attack it fortified by the pooled knowledge of all his predecessors. No elementary textbook contains the principles [to guide the mechanician] ; no one can learn its history." But, he felt, this state of affairs is as it should be. The mechanician and the scientist move in different worlds. For the mechanician, "the workshops, the machines themselves, show what has been achieved, but results depend on individual effort. To understand a machine it has to be divined. This is the reason why talent for mechanics is so rare, and can so easily go astray, and this is why it is hardly ever manifested without that boldness and with the errors which, in the infancy of science, characterize genius."

Evans never rued his deficiency in scientific knowledge but, being a practical man, only his lack of training in mechanical drawing. He began each project with rough pencil sketches, then moved to a chalkboard to reproduce his design in full scale with a two-foot rule, straight-edge square, and compass. He was convinced, probably rightly, that he had lost the chance to build the Philadelphia waterworks because Benjamin Latrobe had sold his version with fine drawings of the dome-shaped pumping house "vomiting its wreath of black smoke that caught the eye of the members of the city council."

"The time will come," Evans mused along lines Latrobe ridiculed as the buggy clattered along, "when people will travel in stages moved by steam engines, from one city to another, almost as fast as birds fly, fifteen or twenty miles in an hour. . . . They . . . travel by night as well as by day ; and the passengers will sleep in these stages."

More than half a century later George Sellers recalled that "Mr. Evans referred frequently to his blasted hopes" during that memorable ride. As Evans had prospered through his Mars Works, it must have been for fame and honor that he had hoped. He died in 1819, aged sixty-four, carrying his blasted hopes to the grave.

8

A Ludicrous Yoking
of Old and New

In 1797 "an unprecedented number of respectable citizens" petitioned the city council of Philadelphia to develop a municipal water system. In 1776 Philadelphia had considered itself the best-watered city in America. Public pumps dotted curbstones everywhere and gave a steady flow of clean, palatable water. But the flow and taste deteriorated as the population surged from about twenty-five thousand in 1776 to more than seventy thousand as the end of the century approached. Housewives after standing in line often received only a trickle of brackish water when they reached the pump. Fire companies complained that the city lacked "a copious supply of water in case of fire." The medical profession, searching for the cause of the string of yellow-fever epidemics that had struck the city annually since 1793, suggested that there might be a connection between the disease and the water supply. The flood of complaints convinced the city council that a crisis had arrived. Something must be done, but what and how? Americans had no experience building municipal water systems.

At this point Benjamin H. Latrobe, a recent immigrant from England, arrived on the scene. He was thirty-four years old, personable, well trained as an engineer and architect, and as luck would have it, had also studied the water systems of several English cities. In 1798 he submitted a plan for a Philadelphia system to the city council in the form of a superbly argued and well-illustrated brochure. His plan called for a steam engine to pump water from the Schuylkill River on the western edge of the city nearly a mile through underground tunnels to Centre Square. There another pump, hidden in a domed building that resembled a Grecian temple, would raise the water to a reservoir, also hidden from view, and from that elevated position it would flow downward and into the city through buried wooden pipes. Nicholas Roosevelt of New Jersey had given assurance that his machine shop could build the required engines. Latrobe estimated that the system would cost $150,000 and take seven months to complete.

Objections were many. A canal company that planned to link the Delaware and Schuylkill rivers complained because it had hoped to profit by feeding water from its canal into the city. Despite Latrobe's assurance that "a steam engine is, at present, as tame and innocent as a clock," many people feared it. There were then only three in the nation, and they appeared strange, awesome beasts to those who had seen them in operation. To critics who observed that the reservoir would hold only 7,500 gallons, a supply the city could exhaust in twenty-five minutes, Latrobe replied that the reservoir could be refilled in six minutes. Oliver Evans, who had a plan of his own, objected loudest. He thought the cost estimate too low, the time allotted too short, and the reservoir barely adequate to supply the city's current needs, let alone those of the future. He argued for the reservoir to be built in the elevated country north of the city that would be large enough to hold many thousands of gallons of water. He called Latrobe's plan a "city plaything on which to expend money, more for ornament than utility."

Members of the city council paid no heed to these objections. They liked Latrobe, a cultivated gentleman like themselves. His credentials were excellent. He was accustomed to supervising complex projects and large gangs of workmen. His plan as presented to them was precise, detailed, and wonderfully illustrated. That report, Evans later admitted, "proved too much for him to overcome." It was those "artistically finished drawings ... of the Boulton & Watt steam engine and pumps, and above all the exterior of the pumping house, with its Doric columns and pediments, both front and rear, its center dome-shaped building covering the reservoir, with the novel expedient of the stack and chimney, terminating on the apex of the dome, vomiting its wreath of black smoke, that caught the eye of the members of the city council.'' Evans was right: the domed temple sold Latrobe's plan. In accepting it, as Charles L. Sanford, a modern scholar, has said, they accepted ''a ludicrous yoking of the old and new and which represents the cultural conservatism of those Americans, especially of the educated class along the Eastern seaboard, who were influenced by inherited traditions of England and Europe.''

Latrobe's domed temple appealed to the council members for another reason. Visitors often complained that Philadelphians thought their city superior ''over every other spot on the globe. All their geese are swans.'' Socially, economically, and intellectually it had long eclipsed Boston and New York, and in 1798 it served as the capital of the new federal government. But it was about to lose this last distinction to Washington, D.C., then being created out of marshland along the Potomac River. Much as the Eiffel Tower three-quarters of a century later served after the humiliating defeat in the Franco-Prussian War to reassert France's preeminence in western civilization, so Latrobe's stately temple might help to retrieve some of Philadelphia's lost prestige. ''Latrobe represented the old cultivated tradition of Europe,'' Charles Sanford remarks, ''while the breed of carpenters, mechanics and tinkerers''—men like Oliver Evans—''belonged to a new vernacular tradition out of which were to come more advanced technical and structural changes.'' Many of the council members must have sensed the impracticability of Latrobe's plan for a growing American city, but the symbolic value of that temple led them to vote for it.

Latrobe up to a point justified the council's faith in him. Construction began early in May 1799 and ended in January 1801, when ''the first water was thrown into the city, about one mile of pipes being laid.'' (This event occurred at the very time that Eli Whitney displayed in Philadelphia his muskets with ''interchangeable parts'' to President Adams and Vice-President Jefferson.) The seven months promised had become eighteen and the $150,000 estimate came closer to $500,000. (Cost overruns were common with Latrobe. Years later, in 1807, when building the capitol and eager to finish the chamber of the House of Representatives, he "ran a deficit of $52,000, about 70 percent over his budget," according to Merrill Peterson. "Jefferson apologized to Congress, and read the architect a stiff lecture on public economy and the principles of American government.") A member of the committee chosen to represent the city's interests during construction came to detest Latrobe—a man "ignorant of his business," he said, "who undertook more than he understood and has been making experiments at the expense of the city"—and to distrust Roosevelt, who overran his budget by $40,000 even before his two engines had been assembled and pumped a drop of water.

Nonetheless, the achievement exceeded anything hitherto attempted in America. Two tunnels six feet in diameter and nearly a mile long had been cut though granite rock and lined in brick to carry river water to the temple. Latrobe, despite the cost overrun, seems to have supervised the huge unique project fairly well. He drew craftsmen and mechanicians from all parts of America

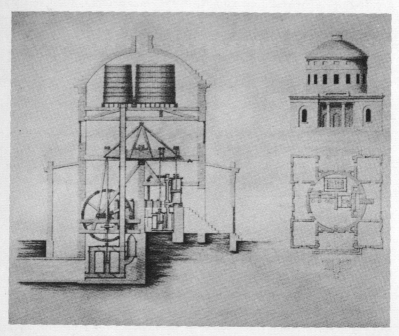

Philadelphia's *Centre Square Water Works with Steam Engine,* an undated ink drawing. *(Historical Society of Pennsylvania)*

and from England, creating a pool of skilled workers who later replicated elsewhere in America what they had done in Philadelphia. (One of his assistants, Frederick Graff, "became the most celebrated waterworks engineer in the United States during his tenure as superintendent of the Philadelphia works.") Nicholas Roosevelt's two gigantic steam engines performed admirably. "The lower engine near Schuylkill is a double steam engine of forty inch cylinder and six feet wide at the bottom, nineteen feet long and ten feet wide at the height of five feet seven inches." Visitors who watched it pump water from the river at twelve strokes a minute knew they had seen a new world coming into being.

Workmen's reluctance to reminisce on paper makes it hard to judge the technological innovations that emerged from building the waterworks. Dr. James Mease, one of the fellow travelers who often visited Coleman Sellers's office after hours, records two. One dealt with Roosevelt's steam engine: "The air pump is an improvement upon that used by Boulton and Watt; consisting in its evacuating the condenser twice at every stroke, thereby creating a much better vacuum, and of course adding considerably to the power of the engine, in proportion to the diameter of its cylinder without increasing friction." The other innovation probably came from English craftsmen experienced in making and laying pipe for a water system. The pipes were of wood, "chiefly of four and a half [feet] and three inches bore," said Mease. They were "bored by placing the logs in two cast iron rings, and centered by regulating screws; as the log turns, the augers enter at each end, and meet in the middle; a pipe of fifteen feet long, can be bored and the joints made for the connecting cylinders in fifteen minutes. The connecting cylinders are of cast iron, widening at both ends, that as the log is driven up the joints become tighter."

The waterworks became the showpiece the city council had hoped for. Etchings of the domed temple abound. It served as a hub for Fourth of July celebrations. When Oliver Evans wanted to demonstrate to the public the steam dredge he had invented, he chose the temple to display it. By then the waterworks had become the plaything Evans had said it was. In 1810 some thirty-five miles of pipe had been laid, but the annual operating deficit was close to $40,000. The reservoir in the temple's dome failed to meet citizens' needs. They resorted to public pumps that had not been closed or ''at night, when the hydrants would run, filled their tubs and buckets for the next day's supply.'' Within fifteen years Latrobe's system was replaced by a large reservoir north of the city like the one Evans had envisioned. Apparently he got no credit for the innovation, but he did have the satisfaction of knowing that one of the steam engines used to pump water from the Schuylkill River into the Fairmount Waterworks was designed and built by Oliver Evans.

9

A Gadget
for the Plain People

Some years ago an American salesman boasted he could sell refrigerators to Eskimos, and he did. More than a century earlier Eli Terry, a Connecticut mechanician, did something similar—he sold mass-produced wooden clocks to thousands of people who had little need for them. Nature dictated the daily routines of farm families. Similarly, working people in villages and towns seldom cared to know the precise time in hours and minutes. "There, too," David Landes has said, "the craftsman awoke with the dawn and the animals and worked as long as natural light or oil lamps permitted. In the typical household workshop, one person, usually the newest apprentice, would 'sleep on one ear,' wake before the others, start the fire, get the water, then get the others up; and the same person would usually shut things down at night." Only the well-to-do in the eighteenth century owned clocks—rarer still pocket watches—which were more social affectations than instruments to order their daily lives. ("The first thing one has to remember," Landes says, "is that it is not 'natural' to want to know the precise time—that is, time as expressed in hours, minutes, and subminutes.") Eli Terry gave the clock to the plain people, regardless of whether they needed it. In 1844 an Englishman traveling in the Midwest said: "Wherever we have been in Kentucky, in Indiana, in Illinois, in Missouri, and ... in every dell of Arkansas, and in cabins where there was not a chair to sit on, there was sure to be a Connecticut clock."

Terry's Thirty-Hour Clock. This simply designed clock was especially popular with peddlers' country customers. It could be mass-produced by unskilled workers, and intense competition among New England clockmakers eventually drove the price down to $7.50 and less. *(Smithsonian Institution)*

Eli Terry was born four years before the Declaration of Independence. He died eighty years later in 1852, the year Harriet Beecher Stowe published *Uncle Tom's Cabin*. He left behind a world he had helped to transform by pioneering the mass production of clocks. He was trained as a boy in the English clockmaking tradition, in which each instrument was made to order. The shops he worked in turned out clocks with both wooden and brass movements. Those of wood were naturally cheaper but unreliable, for the wood often warped, swelled, or shrank, depending on the weather. Brass clocks were more expensive but considerably more accurate. Terry moved to Plymouth, Connecticut, about 1793 and set up in business for himself. Possibly because he saw a slim market among country people in the area, he chose to concentrate on building clocks with wooden movements which he could sell at half the price of those with brass parts. The only brass he had to bring in from outside his shop went into the weights, the pendulum bob, and the crown wheel. His shop held two primitive machines—one, turned by hand, cut the wooden wheels and pinions (cogwheels that meshed with cogwheels); the other was a lathe driven by a treadle.

In the decade that followed Terry must have sensed that a clock was more than a timepiece, that it helped to give its owner social standing. Fernand Braudel remarks that "nothing ever has been effective against the passion to move up in the world." Colonial leaders had sought to keep citizens in their place by imposing sumptuary legislation, but every effort to dictate social status by people's apparel failed. A clocks defined one's social standing in the community as visibly as, say, the wig only gentlemen were supposed to wear. If a clock could be made cheap enough for a modest farmer to afford, he and his wife could feel they had moved up in the world every time they listened to its ticking on the mantelpiece.

Sometime after 1800 Terry took over an old mill and designed machinery to mass-produce wooden clock movements. Nothing is known about the machinery except that what he had previously done by hand and foot was now powered by water. In 1802 he produced more than two hundred clocks, piled them into a wagon, and hawked them about the countryside, selling them for about $30 each. (Later he got the price down to $15 with case, $10 without.) Five years later two brothers, Edward and Levi Porter, who ran a company that finished and assembled clock movements made by others, asked if he could produce four thousand movements in three years. Terry said he could. He spent the first year improving his machinery. In the second he turned out about a thousand movements with the help of twelve workers, and in the third, three thousand. "It was this factory that showed for the first time that mass production of timepieces was feasible and profitable," Landes remarks. "It was also the training ground for such other leaders of the clock industry as Seth Thomas and Silas Hoadley. Terry played the role of teacher to the trade, both in product design and production changes, just as Maudslay did for machine-tool manufacture in Britain." Product design altered in

1814, when Terry introduced a new clock. His first clocks had been the tall, formal affairs long made by those craftsmen; his new model "was both a stylistic and a technological innovation," Hindle notes. "A thirty-hour shelf clock of simplified construction, it could be manufactured by unskilled workers and required less fitting to assemble." A later Terry model, marketed in 1822, became "the standard of the industry for the next fifteen years." Workshops that produced wooden clocks soon proliferated in Connecticut. The capital investment in a shop was small—$2,000 or less—and often much of it came from merchants who distributed the finished product.

Connecticut alone probably could not have absorbed even the four thousand clocks Terry had produced in two years, but it had a distribution system that dated back to colonial days, a network of Yankee peddlers who opened up the continent as a market. Local merchants loaded the peddlers' wagons with Connecticut-made products—brooms and tinware, buttons and pins, nails and nostrums. Every peddler had his own territory and knew most of

Yankee Peddler. The jingling, jangling sound of a peddler's approaching cart, filled with pots and pans, wooden clocks, brooms and baskets, and scores of other "notions," marked a high point in a farm family's drab life. "If we can imagine ourselves, today," Roger Burlingame writes, "growing crops in an isolated spot where no store or village existed, having a five-and-ten on wheels drive up to our door, we may understand the significance of the peddler." *(Charles Scribner's Sons)*

the families along the routes he traveled. Customers generally welcomed him, for he passed along news and gossip with the wares he sold. Naturally, he had a gift of gab, as a foreign traveler observed one day when a peddler, after considerable palaver with a farmer, eased in his pitch.

"I guess I shall have to sell you a clock before I go."

"I expect a clock's of no use here; besides, I ha'nt got no money to pay for one."

"Oh, a clock's fine company here in the woods; why you couldn't live without one after you'd had one awhile, and you can pay for it some other time."

"I calculate you'll find I ain't a-going to take one."

The peddler now turns to the wife. "Well, mistress, your husband won't take a clock. . . . I suppose, however, you've no objection to my nailing one up here, till I come back in a month or so. I'm sure you'll take care of it, and I shall charge you nothing for the use of it at any rate."

When the peddler returns a month later, he finds the whole family charmed with the clock, but the weather had changed, the parts had swollen. It had stopped. The peddler gives the family a new one. Back on the road he tinkers with the one just returned until he gets it running, then sells it to the farmer at his next stop.

The technology Terry created to mass-produce cheap wooden clocks would have come to little without the backing of Connecticut entrepreneurs and their well-developed marketing system to dispose of them. But how does one sell a gadget for which the plain people have no need and which will have little effect on their daily lives? The farmer will still rise with the sun and end the day of work when darkness descends, regardless of what the clock tells him. Here the persuasive peddler entered in. He charmed the wife with the notion that a clock on the mantel would give the family social prestige, and he overcame the husband's resistance ("I ha'nt got no money to pay for one") with a seductive offer—buy now and pay later.

The stream of clocks from Terry's shop came as Robert Fulton's steamboat began to make regular trips up and down the Hudson River, inaugurating a revolution in water transportation. A sailboat moved away from the dock when wind and tide were right and reached its next port at the will of the winds and weather met along the way. Natural elements affected steamboats much less. A public schedule announced the time of departure and the time of arrival, and more often than not the schedule was followed. With the swift spread of the steamboat's presence along America's major rivers, the clock ceased to be a gadget. It shaped daily routines. Shippers, passengers, dock workers, anyone remotely connected with river life depended on it. ("The clock," Lewis Mumford has said, "is not merely a means of keeping track of the hours, but of synchronizing the actions of men.") But Terry's undependable wooden clocks would no longer do. Fortunately, clockmakers following Terry's lead soon began to mass-produce clocks with brass works that were cheap and reasonably accurate. With their innovations clocks began to dominate men's lives in America.

10

Steamboats in the East

With less than his usual restraint, Henry Adams in his great history of the early United States writes that the date that "separated the colonial from the independent stage of growth" of America was 17 August 1807, "for on that day, at one o'clock in the afternoon, the steamboat *Claremont,* with Robert Fulton in command, started on her first voyage.... The problem of steam navigation, so far as it applied to rivers and harbors was settled, and for the first time America could consider herself mistress of her vast resources." This extravagant praise presents a puzzle, for twenty years earlier, almost to the day, John Fitch had launched at Philadelphia while the Constitutional Convention was meeting "not only the first steamboat to move consistently on American waters," James Thomas Flexner remarks in his superb history *Steamboats Come True,* "but the most efficient steamboat that ever had been built by man." In 1790 the improved boat carried out for an entire summer a schedule run between Burlington and Trenton, covering two thousand miles with almost no breakdowns. Yet by the end of that summer he had tied the boat to a wharf and let the hull begin to rot away. Fitch could never duplicate his first success. He died in 1798 bitter and unheralded, and Fulton reaped the glory Fitch had yearned to enjoy. Despite Fitch's cries from the grave, few today deny that Robert Fulton invented the steamboat, even though his boat, like Oliver Evans's automated grist mill, did not contain a single patentable item. His achievement, however, raises a question: Why did he succeed where Fitch, an ingenious gentleman, failed?

A model of John Fitch's boat, 1785. *(New-York Historical Society)*

To understand Fulton's achievement and what is involved in successful technological innovation, let us look at two other mechanical creations, one primitive, the other sophisticated. The first are the farm wagons George Sturt built in his shop in rural England. To modern eyes they seem crude and dumpy. Sturt saw instead "a thing of beauty, comparable to a fiddle or a boat, ... an organism in which all the parts integrated ... reflecting in every curve and dimension some special need of its own countryside, or, perhaps, some special difficulty attending wheelwrights with local timber." Now a jump to modern times. There would seem to be no resemblance between a modern commercial airplane and one of Sturt's wagons, built by hand tools that "took us no nearer to exactness than the sixteenth of an inch." Yet what held for his craftsmen—"necessity gave the law to every detail"— holds for the builder of the airplane, which John Newhouse has described as "a bewildering mixture of exquisitely balanced trade-offs." The speed desired, the distance to be covered, the load to be handled—these and scores of other give-and-takes must be integrated in the final product.

Fulton's achievement lies midway between the wagons and the airplane. He succeeded where a baker's dozen of other steamboat pioneers failed because he recognized, confronted, and solved the dilemmas of "exquisitely balanced trade-offs" while building his boat. Fitch neither saw nor comprehended the dilemmas, as Flexner notes: "Approaching a machine-age problem from the point of view and with the tools of an eighteenth century craftsman, he had no conception of the vast number of elements that would have to be co-ordinated before a steamboat would function efficiently. He did not realize that in addition to forms, you had to deal with stresses and proportions and the properties of materials.'' Fulton —probably because, as Brooke Hindle has argued, his training as a painter taught him to visualize a complex creation as a unified whole—had what Fitch lacked. Later, when accused of stealing ideas from others, he admitted that his boat, like Evans's auto-mated mill, contained no patentable items, also that he had stud-ied all Fitch's plans and diagrams. ''Every artist, who invents a new and useful machine, must choose it of known parts of *other* machines,'' he said, ''I made use of all these parts to express ideas of a whole combination, new in mechanics, producing a new and desired effect, giving them their powers and proportions indis-pensable to their present success in constructing steamboats.'' For instance, he went on, he had studied the tables of Charnock, a naval architect, on water resistance.

> I drew from those tables such conclusions as ... to ascertain as near as possible the resistance of any given boat, and from her resistance also shew what would be the power of the steam en-gine to drive her the required velocity, then shew what should be the size of the wheel boards, which take the purchase of the water, and their speed compared to the speed of the boat, all of which were necessary to be ascertained, selected and combined before any one could originate a useful steam-boat; and it was for want of such selection and just combination of first princi-ples, founded on the laws of nature, that every attempt at con-structing useful steam-boats *previous to mine failed*.

Fulton's conception of the steamboat as a creation of ''balanced trade-offs'' does not alone explain his success. Personality enters in to an extent. Fitch counted among the dirty-fingernail people of the day, which helps to make clear his lack of appeal in winning support from political, social, and intellectual leaders. He rubbed people the wrong way. Fulton, by contrast, was a song-and-dance man as adept and smooth as Whitney. While in Europe he sold under false pretenses the rights to a rope-making machine in-vented by an Englishman. He copied all the basic ideas of David Bushnell's submarine and convinced the French ''that every as-pect of submarine warfare was altogether original with him.''

These shenanigans apparently did no harm. He continued to move comfortably among the elite in America and Europe. When Chancellor Robert R. Livingston, one of New York's most eminent and wealthiest gentlemen, came to Paris to negotiate the Louisiana Purchase, he met Fulton and in no time at all agreed to help subsidize the steamboat Fulton wanted to build.

An agreeable personality only in a small way explains Fulton's success. James Rumsey, one of Fitch's chief competitors, was an equally ingratiating person. Franklin and Washington backed him with enthusiasm. Jefferson thought him "the most original and the greatest mechanical genius I have ever seen." Friends raised money to send him to England to study mechanical advances there and to buy a Boulton and Watt engine. But Fitch, the loner, in a constant scramble for financial support, beat out the affable Rumsey with the first workable steamboat. Rumsey lost mainly because he rejected Fitch's plan to power a boat by paddle wheels and accepted Franklin's proposal of jet propulsion —sucking water in through the prow of the boat and using the steam engine to propel it out the stern. So much for personality.

Andrew Carnegie once said, "Pioneering don't pay." This remark, more than dirty fingernails and an abrasive character, helps to explain why Fitch failed: he was ahead of the times. Philadelphia in his day had few mechanicians, and he seems seldom to have mingled with even those few. He had an ingenious assistant in Henry Voight, a German immigrant—"the first me-

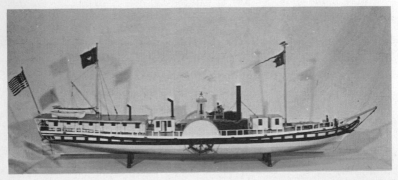

A model of Fulton's *Chancellor Livingston*, 1815. *(Smithsonian Institution)*

chanical genius I ever met with," said Fitch—but when Voight left he had to push ahead largely on his own. "The skills needed for executing the best of steamboat plans did not exist in America," when Fitch was working on his boat, Brooke Hindle has written. "Clockmakers were the finest mechanics available, but the tolerances they achieved were less than those required in the working of a steam engine. Besides, they did not work with the same kind of materials, and the blacksmiths and brass founders who did had little experience with anything but gross tolerances." All of these factors had changed twenty years later. Fulton had a community of excellent mechanicians in the New York area to call on for advice and help, and when these people failed him he could entice such outsiders as Paul Revere, who came down from Massachusetts to install a copper boiler on the *Claremont*.

The sedate launching of the *Claremont* differed sharply from one that Brissot de Warville witnessed of Fitch's boat in 1788. Brissot doubted that the steamboat would ever be "commercially useful," but he bristled at the hoots from the crowd as Fitch's creation chugged away from the dock. "I could not suppress my indignation upon seeing Americans frustrate and discourage with sarcastic jeers the noble effors of one of their fellow citizens. When will men ever come to each other's help and encourage one another with real assistance, rather than discourage one another with ridicule?" The crowd's reaction resembled that of Americans a century later when they passed a Model T Ford mired in mud and yelled, "Get a horse." Mockery helps to disguise sharp breaks with the past that engender fear. No jeers greeted the *Claremont*'s departure up the Hudson, partly because as the years had passed the public had learned to live with the steam engine. Nicholas Roosevelt's two monsters in the Philadelphia waterworks had hissed and rumbled for six years without accident. The scene at the launching site, Flexner writers, "might have been a fashionable picnic that was gathering there. Each man was elegant in spotless ruffles and professionally arranged hair; each lady dimpled charmingly from under a correct bonnet, from over a stylish dress. The Livingston clan, the most mighty political and social force in New York State were gathering with their friends and hangers-on to countenance the adventure of their patriarchal leader, the Chancellor."

Finally, Fitch failed because he chose the wrong river on which to experiment with his boat. This fact seems paradoxical, for he, like Fulton, did not build his steamboat for eastern rivers. They designed them "with a view to the navigation of the Mississippi from New Orleans upward." Both men knew they needed time to perfect their boats for western waters, which moved too swiftly for sailing vessels to make headway. They also knew their boats could not compete in the East with sailing vessels, which got their power free from the wind. The huge low-pressure engines were costly to operate and left little room to carry bulky goods. Passenger service alone offered a way to bring in enough cash to finance further experiments. Fitch tried that route and failed because the distance between Philadelphia and the falls at Trenton, about forty miles, was too short to attract customers who could make the brief trip by horseback or wagon at little cost and in about the time it took the steamboat. Fulton also tried the passenger route and made it profitable because the Hudson offered a clear way for more than 150 miles from New York to Albany with scores of villages along the way where travelers could be landed and picked up. (The charge was five cents a mile or about $7.50 for a ride from New York City to Albany.) Each year he reinvested the profits in a new, enlarged, and improved boat, and slowly, with an eye to attracting more passengers, the boat evolved "into a floating palace, gay with ornamental paintings, gilding and polished woods."

Our Philadelphia snob, Sidney George Fisher, describes a trip in 1847 on one of these floating palaces.

Went to Newport on Thursday afternoon at 5 in the *Bay State*, another wonder of modern art, built since last year. She is of immense size and fitted up in the most complete & costly manner. The machinery is bright as silver & exhibited thro large panes of plate glass set in gilded frames. The saloons, cabins & staterooms are all painted & gilded in the most splendid style & sumptuously furnished. Brilliant Saxony carpets, chandeliers, marble tables, sofas, ottomans, armchairs of every pattern, well-cushioned & covered with the richest stuffs, silk curtains, French china, cut glass, mirrors, fill every apartment. The table is excellent and all the apparatus of the best kind, a napkin &

silver fork to every plate, the servants are well dressed & well drilled, & order, cleanliness & comfort reign throughout. It seems impossible to add anything to this magnificent steamer, tho no doubt another year will see one produced with many improvements & new embellishments. The vast multitudes, increasing every year, who throng the great routes of travel sustain & justify this lavish expenditure whilst competition produces all this comfort & lowers the price. The fare to Newport is only $1 without a stateroom & $3 with one. I had one to myself by applying the day before & slept as comfortably as if in my own chamber. The berths, however, are so good that except in hot weather there is no great difference.

In 1810, five years before Fulton died, Nicholas Roosevelt moved to Pittsburgh to set up a machine shop to build engines that the Fulton-Livingston monopoly hoped would soon launch a fleet of steamboats on western waters. Only now would Henry Adams's prophecy that the steamboat would let America become ''mistress of her vast resources'' begin to be fulfilled.

11

Steamboats
on Western Waters

The story of steamboats on western waters begins with Oliver Evans, that luminous presence among early mechanicians. Evans believed he deserved at least some credit for inventing the steamboat. He had, he said toward the end of his life, "discovered the powerful principles of my [high-pressure] engines" in 1775, and as early as 1781 "I began to try to induce people to apply them to boats on the western waters." He knew that his engine consumed more fuel than the Boulton and Watt but was more powerful, and at a time when notions were vague about rivers west of the Appalachians, he knew that it would be ideal for the western rivers, which were too narrow, swift, and twisting to accommodate sailboats moving upstream. "Thousands of spectators" viewed the steam dredge he launched in 1804, but Evans put the monster in storage and did not follow up his achievement, possibly because he had to earn a living and, like Fitch and other dirty-fingernail people, he could not round up enough capital to pursue his creation. Fulton's tenacity and Livingston's money led three years later to the *Claremont*'s successful trips up and down the Hudson. The glory went to Fulton, but Evans, though he remains little more than a footnote in history books, conquered the western waters with his high-pressure engine. Even there, however, the fame he yearned for never came.

In 1810 Nicholas Roosevelt moved from his New Jersey machine shop to Pittsburgh, and with backing from Fulton and Livingston and help from eastern mechanicians built an engine for the *New Orleans,* the first steamboat on western waters. It descended the Ohio and Mississippi rivers in 1811. Roosevelt saw at once that the history of steamboating had entered a new chapter. Half-submerged trees—called snags, sawyers, or planters—that had tumbled from the soft riverbanks threatened to rip apart his frail hull. He knew the underpowered Boulton and Watt engine would never make it back upstream against the swift currents of the upper reaches of the rivers. After arriving at New Orleans, he confined the boat's scheduled runs to the lower Mississippi. There a denser population offered profitable traffic, and there the broad snag-free river was easier to navigate.

While the *New Orleans* was inching its way downstream, Oliver Evans set western steamboating on a new course. In 1811 he sent his son George to Pittsburgh to start building high-pressure engines for western waters. The engine was compact, relatively simple to build and repair, and according to Evans, ten times more powerful than the Boulton and Watt. It weighed less than five tons, compared with more than one hundred tons for the English engine, and thus occupied a twentieth of the space, leaving plenty of room for freight as well as passengers. True, it burned huge amounts of wood, but the heavily forested riverbanks made this drawback of little importance. Full credit for this gift to the West never redounded to Evans, partly because his own manual, *Abortion of the Young Steam Engineer's Guide,* gave such detailed instructions that any literate mechanician who followed them and had tools and materials available could produce his own version. Within a decade Evans's original model had been modified into something new. It was still a high-pressure engine, but alterations had also made it into, as Louis Hunter puts it, a "noncondensing, direct-acting, horizontal-cylinder affair with cam-action gear." This technical description may mean little to a layman, but it makes the point that a host of anonymous "artists" had by the 1820s created a new machine suited to the needs of steamboating on western waters.

The same holds true for the boat itself. The first ones on western rivers, built by eastern workmen, resembled seagoing vessels; some even carried masts and sails. The deep hull with its thick planking, the weighted keel, the sheered bow and stern, were designed for running through heavy seas. Builders put the steering mechanism in the stern, as on sailing ships. Within two decades river men transformed the boat into something totally new. First to go were the heavy hull and keel, supplanted by a flat, raftlike platform that drew little water. (The western steamboat, said one man, is "an engine on a raft with $11,000 worth of jig-saw work.") The western steamboat, said an early admirer, "must be so built that when the river is low and the sandbars come out for air, the first mate can tap a keg of beer and run the boat four miles on the suds." Everything on the boat—engine, fuel supply, freight, passenger accommodations—was placed above deck. The steering wheel was moved forward to a cabin above the superstructure that gave an unobstructed view of the river. The boat was built of thin planking tied together with "tin, shingles, canvas, and twine." It looked and was fragile yet durable enough for river travel. "If a steamboat should go to sea," a contemporary said, "the ocean would take one playful slap at it, and people would be picking up kindling on the beach for the next eleven years." Rickety as they were, western boats inspired awe among visiting Europeans. "The western steamboats . . . are huge houses of two stories," said a Frenchman. "Two large chimneys of columnar form vomit forth torrents of smoke and thousands of sparks; from a third a whitish cloud breaks forth with a loud noise; this is the steam-pipe. In the interior they have that coquettish air that characterises American vessles in general; the cabins are showily furnished and make a very pretty appearance."

Though outwardly much alike, western steamboats failed to find a fixed form until the eve of the Civil War. Constant tinkering by boatbuilders and mechanicians led to an endless stream of innovations. Some builders preferred the paddlewheel in the stern. Others stayed with sidewheels but to improve maneuverability installed separate engines to power the wheels. Still others, their eyes on the profitable passenger trade, concentrated on making the boats "very fine and comfortable." The result, said a visiting Englishman, was that "no two steamboats are alike, and few of them have attained the age of six months without undergoing some material alteration." The normal life of a steamboat was four or five years, but one captain told "of a boat about the construction of which he had taken great pains," yet "she died at three years," an obsolete hulk.

The boats' short lives did not deter investors. In 1818 there were twenty steamboats on western waters, two years later there were at least sixty, and by the end of the decade more than two hundred were making regular runs. The cost of an average boat, about $20,000, could be paid off in a round trip or two, and by the time it reached senility the owners had taken in a small fortune. Even the largest, plushest boat, "notwithstanding . . . the elegance, . . . including their engines and furniture," said a Frenchman, cost no more than $40,000. In France, he added, a similar boat would cost about $100,000, mainly because of the high cost of wood and of skilled mechanicians. "The Americans," he said, "excel in working in wood."

A trip up or down western waters held risks for every passenger. The boats were not technologically deficient, but their firemen were hellbent to outrace every craft on the river, and to that end taped down the safety valve on the boilers of their high-pressure engines. The accident rate was appalling. In the early years of western river travel 150 explosions killed some 1,400 people. Boorstin has brought from the past the comment of a western riverboat fireman to an eastern traveler in 1844 on the awful accident rate. "Talk about *Northern* steamers," he said, "it don't need no spunk to navigate them waters. You haint bust a boiler for five years. But I tell you, stranger, it takes a man to ride one of these half alligater boats, head on a snag, high pressure, valve soddered down, 600 souls on board and in danger of going to the devil."

Few individuals loom from the misty past in the transformation of Fulton's creation on western waters into something new. Daniel French, a mechanician, is credited with modifying Evans's machine into a compact, direct-acting engine on a stern paddlewheel. Major Stephen H. Long of the Army Corps of Engineers added the "cam-cutoff" to increase efficiency, and also as the corps's superintendent of western river improvements designed the first effective boat to remove obstructions from the rivers. In most histories Henry M. Shreve emerges as the hero of steamboating in the West. He flouted the monopoly Congress had granted Fulton and Livingston on western waters, and later, as

Long's successor as superintendent—Shreve became the first civilian superintendent of western river improvement—his snag
boat, among other achievements, removed the infamous Red River
Raft.

The word "snag" covers a multitude of the Mississippi's sins
imposed on rivermen. John McPhee has described the variety with
his usual accuracy and vividness. The worst "were huge trees that
had drifted south over the years and become stuck in various
ways. One kind was rigid in the riverbed and stood up like a
spear. It was called a planter. Another, known as a sawyer, sawed
up and down with the vagaries of the current, and was likely to
rise suddenly in the path of a boat and destroy it." When these
snags collected into a mass, they were called a "raft." The Red
River Raft held in bondage tens of thousands of uprooted trees
and stretched for 160 miles. Shreve's snag boat cleared eighty
miles of the raft in one year and went on to finish the job later.

Louis Hunter is disinclined to accept any part of the central
role other historians have given Shreve. "The suggestion of the
method of attacking the Raft of the Red River, which proved
successful," he says, "originated in the Office of the Chief of
Engineers." Moreover, for "many of the contributions to structural and mechanical development of the western steamboat" attributed to Shreve, "the meager contemporary evidence supplies
neither confirmation nor contradiction." Hunter holds that "if
the returns were all in," the achievements of Shreve, along with
Long, French, and others, "would assume a quite modest position
beside the collective contribution of scores of master mechanics,
ship carpenters, and shop foremen in whose hands the detailed
work of construction, adaptation, and innovation largely rested."
The heroes of steamboating on western waters "are the anonymous and unheroic craftsmen . . . in whose hands rested the daily
job of making things go and making them go a little better."
"The story of the evolution of steamboat machinery," Hunter
says elsewhere,

in the end resolves itself in large part into such seemingly small
matters as, for instance, machining a shaft to hundredths instead of sixteenths of an inch, or devising a cylinder packing
which would increase the effective pressure a few pounds, or

altering the design of a boiler so that cleaning could be accomplished in three hours instead of six and would be necessary only every other, instead of every trip. Matters such as these do not often get into the historical record, yet they are the stuff of which mechanical progress is made, and they cannot be ignored simply because we know so little about them.

In other words, proceed cautiously in any account of the nuts and bolts of the American past.

12

A Wave of the Future

Thomas Jefferson in his *Notes on the State of Virginia,* published in 1785, remarked that America would do well to avoid the industrial blight that was overtaking England and confine herself to the virtuous pursuit of agriculture. He could tolerate within the "compass of a private family" such machines as a spinning jenny, a thresher, or an automated flour mill, but not the dismal factories he had seen in England that blotted the sky with smoke and debilitated the men, women, and children who worked in them.

The War of 1812 had just ended when an acquaintance reminded him of these early views. "You tell me I am quoted by those who wish to continue our dependence on England for manufactures," he replied. "There was a time when I might have been so quoted with more candor, but within the thirty years which have since elapsed, how are circumstances changed!" Generalizations, he went on, "will depend on the circumstances which shall then exist; for in so complicated a science as political economy, no one axiom can be laid down as wise and expedient for all times and circumstances." England for a generation had made a mockery of America's belief that "free ships make free goods" and had harassed her commerce with "profligacy and power enough to exclude us from the field of interchange with other nations." She had left the United States no choice but to manufacture for herself. "He, therefore, who is now against domestic manufacture, must be for reducing us either to dependence on that foreign nation, or to be clothed in skins, and to live like wild beasts in dens and caverns. I am not one of these; experience has taught me that manufactures are now as necessary to our independence as to our comforts.''

Jefferson had convinced himself he could accept the numerous woolen and cotton mills that had proliferated during the War of 1812. Clean, smokeless water powered them and most were small enough that the number of employees barely exceeded the ''compass of a private family.'' It is fair to wonder, however, if he would have endorsed an experiment going on in Massachusetts and of which he was probably only dimly aware—the creation in 1814 of America's first large textile factory at Waltham, and a later, larger complex at Lowell, both inspired by Francis Cabot Lowell.

Lowell's achievements in a short life—he died in 1817 at the age of forty-two—were remarkable and deserve the attention historians of early American industry have lavished on them. He was blessed with brains and money, along with a bold but disciplined imagination. He can be ranked as the preeminent fellow traveler of his day because he also had the good sense to rely heavily on one of the most distinguished among the dirty-fingernail people in New England. In 1858 Nathan Appleton, then seventy-nine and nearing the end of a long life, reminisced about the enterprises Lowell had inspired. Early in his recollections he remarks, ''The first measure was to secure the services of Paul Moody, of Amesbury, whose skill as a mechanic was well known, and whose success fully justified the choice.'' Moody was thirty-four when Lowell drew him to Waltham. He had worked in English and American mills and with Jacob Perkins, who had patented the first nail-making machine, yet he knew as he heard Lowell's plans that his ingenuity as a mechanician had never been taxed as it was to be in the future.

Jefferson's Embargo of 1807 had put a damper on American commerce and Lowell's prosperous mercantile business. The approach of the War of 1812 convinced him that the English would find it more and more difficult to export their mass-produced cloth to America, and during a leisurely holiday in England he studied with his usual thoroughness its textile industry. The sight of the power loom, then being introduced into the industry, convinced him that a factory in America built around such a loom would prosper in the gap left by the absence of British competition. But his factory must differ from the dismal piles of brick he had seen in cities like Manchester, where the machines were manned by pale-faced, often emaciated and drunken employees who seldom saw the sun through the smog-filled sky.

By the time Moody arrived at Waltham, Lowell had formed in his mind the factory he wanted, adapted to American conditions. It would be powered by cheap, clean water. It would be located in the country and festooned with large windows that would let in light but also allow the workers to look out on a bucolic scene. To avoid rearing up a permanently depressed working class, he would create a pool of "circulating" workers—young ladies drawn from the New England countryside. They would work for half the pay of men, but Lowell offered them the chance to exchange the drudgery of farm life for a salary, however small, to live in a distant, almost exotic community, and the promise that they could return home whenever they tired of factory life. Another promise reassured hesitant parents that the young ladies' moral lives would be closely supervised and that they would depart from Waltham as pure as when they had arrived.

Power-loom weaving in the Lowell Factory, 1835.

Lowell's complex vision for his factory had three further aspects to it. First, it would produce a single product for a mass market, a cheap, sturdy cloth or sheeting, wide enough that housewives could cover a bed with it or convert it to a variety of other uses. Second, his mill, unlike those in England, would be fully integrated. Raw, baled cotton would come in on the ground floor. After it had been picked—the matted fibers shaken loose—and carded, it moved to the second floor to be spun into yarn, then to the top floor where the power looms wove it. Mr. Lowell, said his friend Appleton, "is unquestionably entitled to the credit of being the first person who arranged all the processes for the conversion of cotton into cloth, within the walls of the same building. It is remarkable how few changes have since been made from the arrangements established by him, in the first mill built at Waltham."

(A New Hampshire millowner some time later summed up the revolution Lowell had worked on the manufacture of cloth in America. "In England," he said, "not only are the machines and manner of operation of them on a different plan, but they have their spinning one place, their dressing another, their weaving scattered through various cottages and cellars, and their bleaching and printing somewhere else." The gentleman, aware of his and his firm's deficiencies, goes on: "One who has been accustomed to attend one branch of the business would hardly be qualified to superintend an establishment here where all the various departments from the picking of the cotton to the baling of the printed goods are carried on within the same yard.")

Finally, because the mill was going to stand in the countryside surrounded by what a foreign visitor called "an extemporaneous town," some distance from a handy network of shops, most of the machinery had to be designed, built, and maintained by its machine shop. Herein lay the need for Paul Moody. He and Lowell made a perfect team, though Moody would have been quick to admit that Lowell "was the informing soul, which gave direction and form to the whole proceeding," and that he "made no alterations or improvements in any part of the works without consulting" him.

An anecdote related by Appleton illustrates how well they worked together. They were visiting Silas Shepard of Taunton, whose patented machines Lowell was "chaffering" to buy the rights to, but at a lower price than Shepard wanted. Shepard turned to Moody and said, "you must have them, you cannot do without them, as you know, Mr. Moody."

"I am just thinking that I can spin the cops direct upon the bobbins," Moody said.

"You be hanged," said Shepard. "Well, I accept your offer."

"No," said Lowell. "It is too late."

Lowell's vision set conditions for Moody no other American mechanician had ever confronted. He must build machines sturdy enough to take a relentless pounding, yet cheap enough to be easily replaced when innovations came along. They must be so simple to operate that a young lady after a few days of training could work alone. As in the song that finds "the shin bone is connected to the thigh bone, the thigh bone is connected to the hip...," so every machine in the mill had to be shaped and adjusted to fit in the assembly line with the one ahead and the one behind it. The power loom, for instance, "necessitated higher operating speeds in the prior processes of warping and dressing," David Jeremy remarks. "Unless these were achieved, power loom weaving would be uneconomic."

The mill opened early in 1815, and within the year Moody had "invented a machine for warping," according to the plant manager. "The next year he made important and useful improvements in the English Dressing machine for sizing yarn." (He never hid the fact that a number of his innovations were improvements on machines invented in and pirated from England.) He also built an improved spinning frame that called for "only five persons to do the same quantity of work that requires at least twenty persons to do and saves much of the cotton which is unavoidably wasted in the usual method and requires much less power, less space, and the cotton comes out much more equal." He designed a "stop motion" gear that shut down an operation when yarn broke or a machine ran out of material to process.

Moody and his men not only kept the mill running, they also contributed substantially to its profits. In one five-year period, machines built in his shop and sold to other mills added nearly $100,000 to the balance sheet. Further profits came from selling rights to patents Lowell had bought up and to inventions that

Moody had created. It is unclear how much he profited personally from his ingenuity, but it is likely that when he teamed up with Lowell he had accepted an agreement similar to one signed by an assistant in 1820: ''Should I be fortunate enough to make or suggest any improvement for which it might be thought proper to obtain a patent such patent or patents are to be the property of the Company.''

Any resentment Moody felt may have been eased when a group of Boston businessmen in 1821 formed the Merrimack Manufacturing Company, initially capitalized at $600,000. They gave Moody sixty shares in the company. He soon moved to Lowell, a new extemporaneous town named after his friend, who had died in 1817. (The village, incidentally, soon became a city. In 1855 fifty-two mills strung along the Merrimack employed 8,800 women and 4,400 men and produced 2.25 million yards of cotton a week.) Moody began a second career in Lowell that lasted until 1831, when he died at the age of fifty-two. Men trained in his shops carried what they had learned to all parts of the country but especially throughout New England. They were a new breed of men. As much as Francis Cabot Lowell, Paul Moody in his way had helped to create a wave of the future. Jefferson, for all his alertness to the ways circumstances had changed during the War of 1812, probably never realized how much they had changed in 1826, when he died only dimly aware of what had been happening up in New England.

III

Guns and Government

The year 1815 was one of jubilation in the United States. The War of 1812 had ended and, amidst bonfires, cannon salutes, and tolling church bells, citizens everywhere joined in a national celebration the likes of which had not been seen since the Revolution. Republicanism had been vindicated, despotism vanquished, and independence finally achieved. Exhilarated by the public optimism and air of self-congratulation that followed the Treaty of Ghent, the popular press conveyed the feeling that a new era of "Peace and Plenty" was at hand. Slowly but perceptibly the Revolutionary ideals of liberty and republican virture were beginning to find renewed meaning in expectations of progress and prosperity. In such a milieu, nothing seemed impossible. Wiser men knew, however, that the country had barely skirted disaster during the war; good fortune rather than military might had ensured the nation's survival. From the beginning, faulty arms, insufficient supplies, and tactical errors had plagued the war effort—facts that no one, in 1815, appreciated more than the Secretary of War and his general staff.

—Merritt Roe Smith

13

A Song-and-Dance Man?

A song-and-dance man is someone who "exaggerates or purposely confuses the facts." Was Eli Whitney a song-and-dance man? Even to raise the question can outrage those wed to old accounts of American technology's past. Whitney invented the cotton gin and, so tradition holds, fathered the American system of using uniform or interchangeable parts. No one today can take the cotton gin away from him, though hundreds of southern planters did to the extent that they pirated his invention and paid him no royalties. Interchangeable parts are another story.

On 14 June 1798, six years after he invented the cotton gin, Whitney signed an extraordinary contract with the federal government. He promised to deliver to the army within a year and a half four thousand assembled muskets. He did so at a time when the government's two armories, one at Springfield, Massachusetts, and the other at Harpers Ferry, Virginia, were turning out no more than a trickle of weapons. Even more remarkable, the government agreed to subsidize Whitney's project. Normally, at that time, private contractors, like the gunsmith Simeon North of Connecticut, were paid only on delivery. And the subsidy went to a man who was close to bankruptcy; lawsuits against those who had pirated his cotton gin had swallowed up the little he had made on the invention. Finally, he knew nothing about making guns. How had Whitney pulled off this ingenious "scam"?

Whitney was an opportunist. In 1798 war with France—the Quasi-War in history books—loomed. The government feared that Napoleon might at any time invade the United States. Troops must be armed, but the inadequate federal armories obviously could not do the job. Whitney saw an unfulfilled need and moved in with his proposal. The political leaders bought it for several reasons. He came to them with good credentials. The ingenious cotton gin had already begun to transform a southern economy in which cotton was soon to be king. The "old school tie" opened doors for him. He was bright, articulate, and personable, a gentleman who had been educated at Yale, able to read Latin and to converse easily with other gentlemen like President John Adams and Vice-President Thomas Jefferson. The secretary of the treasury, Oliver Wolcott, also a Yale graduate and also from Connecticut, backed Whitney from the beginning and supported him long after others suspected they had been hypnotized by a song-and-dance man. (It took ten years for Whitney to complete his original contract.) But it called for more than political pull and

Eli Whitney (1765–1825). *(New York Public Library)*

a glib tongue to gain the backing of the astute leaders of the early republic. He had come to them with a new idea—a complicated mechanism like a gun need not be constructed by skilled craftsmen. Machines, he said, can do the job.

''I am persuaded,'' Whitney told the government, ''that machinery moved by water, adopted to this business, would greatly diminish the labor and facilitate the manufacture of this article.'' Jefferson, an admirer, found nothing to object to in this remark, but he might have if he had heard Whitney expand on the implications buried within it. ''My intention is to employ steady, sober people,'' Whitney said at another time, ''and learn them the business. I shall make it a point to employ persons who have family connections [i.e., married men with children] and perhaps some little property to fix them in place, who consequently cannot be easily removed to any considerable distance.'' He would, in short, make the workers captives to his workshop and thus solve the problem of a frequent turnover of labor that plagued all who dealt with peripatetic mechanicians.

He would further tighten the hold on workers with a revolutionary idea that would have appalled Jefferson : he would take

The Cotton Gin. These two pictures are of an early demonstration model of Whitney's machine. Hindle and Lubar describe how it worked : ''The hand crank moved the cotton to the sawtooth wheels, which pulled the fiber through the wire slots, separating it from the seeds that fell to the bottom of the gin. The function of the brushes was to move the cotton and to clean it off the sawteeth.'' *(Smithsonian Institution)*

the artist out of the artisan. He said as much when he proposed "to substitute correct and effective operations of machinery for that skill of the artist which is acquired only by long experience." Before this time the mechanician had combined mental with manual work. "Nathan," the Philadelphia mechanician had said, "when I hire a workman I hire his brains as well as his hands." Whitney would have none of that blather. He would hire inexperienced men with nothing to unlearn. He would teach them no craft, no skill. They would only man a particular machine that made a particular part. If the worker chose to quit, he departed knowing little more about the mechanician's world than when he arrived, but he left behind the lathe, the drop hammer, or whatever machine he had operated, a machine that any normal lad could learn to run in a week or so. Some thirty years later Alexis de Tocqueville commented on the implications of Whitney's vision, which Thomas Jefferson had missed until it was too late:

> When a workman is unceasingly and exclusively engaged in the fabrication of one thing, he ultimately does his work with singular dexterity, but at the same time he loses the general faculty of applying his mind to the direction of the work.... What can be expected of a man who has spent twenty years of his life in making heads for pins?
>
> Whereas the workman concentrates his faculties more and more upon the study of a single detail, the master surveys a more extensive whole, and the mind of the latter is enlarged in proportion as that of the former is narrowed. In a short time the one will require nothing but physical strength without intelligence; the other stands in need of science, and almost genius, to insure success. This man resembles more and more the administrator of a vast empire—that man more a brute.
>
> The master and the workman have then here no similarity, and their differences increase every day.... Each of them fills the station which is made for him, and out of which he does not get: the one is continually, closely, and necessarily dependent upon the other, and seems as much born to obey as that other is to command. What is this but aristocracy?

Let us forget for a time the implications of Whitney's first vision and move on to his second vision—the production by machinery of uniform or interchangeable parts for the guns he had promised the government. This plan came slowly to Whitney, about a year after signing his contract, and it did not originate with him. It came either directly or in a roundabout way from Thomas Jefferson. In 1785, Jefferson had visited the workshop of a French mechanician, Honoré Blanc, and soon afterward he described to a friend back home what he had seen.

> An improvement is made here in the construction of the musket which it may be interesting to Congress to know, should they at any time propose to procure any. It consists in making every part of them so exactly alike that what belongs to any one may be used for every other musket in the magazine. The government here has examined and approved the method and is establishing a large manufactory for this purpose. As yet the inventor had completed the lock [the assembled parts of the trigger mechanism] only of the musket on this plan. He will proceed immediately to have the barrel, stock and their parts executed in the same way. Supposing it might be useful to the U.S. I went to the workman. He presented me with the parts of 50 locks taken to pieces and arranged in compartments. I put several together myself, taking pieces at hazard as they came to hand, and they fitted in the most perfect manner. The advantages of this, when arms need repair, are evident.

At the end of 1800 Whitney had delivered few of the four thousand muskets promised. He needed funds and a dramatic event to assure that the flow of money from the federal treasury continued. In January 1801, before an audience that included President John Adams, President-elect Thomas Jefferson, and a number of other leading politicians, he duplicated the show Blanc had given Jefferson five years earlier. Everyone left impressed, especially Jefferson. "Mr. Whitney," he reported, "has invented moulds and machines for making all the pieces of his locks so exactly equal, that take 100 locks to pieces and mingle their parts and the hundred locks may be put together as well as by taking the first pieces which come to hand." From that day forward Whitney became a new kind of American hero—"Artist of his Country." So he remained until modern scholars became curious. Merritt Roe Smith has summarized their findings:

Whitney's reputation as inventor of the American System stood
unchallenged until the 1960s when researchers began to detect
serious discrepancies between the written record and extant ar-
tifacts. Most damaging was the discovery that Whitney's muskets ... did not have interchangeable parts. The physical
evidence was unmistakable ... that Whitney must have staged
his famous 1801 demonstration with specimens specially prepared for the occasion. In short, it appears that Whitney purposely duped government authorities in 1801 and afterwards
encouraged the notion that he had successfully developed a system for producing uniform parts.

Obviously, Whitney was a song-and-dance man. Was he a char-
latan? Contemporaries who had worked with him did not think
so. The War Department continued to call on him for advice.
Roswell Lee, an early assistant who later became superintendent
of the Springfield armory, remained a friend and admirer and at
the armory refined and elaborated Whitney's two visions—uniform, machine-made parts of a complex instrument. Let Harold
Livesay's judgment of Eli Whitney stand for the time being:

Among these pathfinders, Whitney ranks first, but not because
he originated interchangeability, perfected it, or applied it successfully to mass production; he did none of these. What he did
do was to perceive identical parts as a precondition to volume
production, particularly in the circumstances of the early
United States. He applied his energy and genius to the problem
and influenced his contemporaries and successors who addressed
the same challenge. Whitney built the first arch in the engineering bridge that spanned the gulf between the ancient world's
handicraft methods of production and the modern world's mass
manufacturing, a bridge across which his countrymen and the
rest of the so-called developed world marched to prosperity.

14

The Springfield Armory

There is little doubt that Whitney was a song-and-dance man, but those scholars who have chipped away at achievements he exaggerated have overlooked an important fact—the respect, even affection, contemporaries had for him. His foreman James Carrington worked with him for twenty-six years and left only when Whitney died. Decius Wadsworth, head of the army's Ordnance Department, often sought his advice and remained a friend over the years. Roswell Lee, an early protégé, stayed in close contact after returning from the War of 1812 as a lieutenant-colonel and becoming superintendent of the government's Springfield Armory. Whitney inspired these gentlemen and others to focus on three concepts he sought to realize in his own workshop but never did—the creation of a stable labor force trained to perform unskilled, routine work; the development of machines that would do what skilled gunsmiths had always done by hand; and the production of precisely machined gun parts that were interchangeable from one weapon to another. Roswell Lee pursued all three goals during his eighteen-year tenure at the Springfield Armory. He achieved the first, succeeded partially with the second, and failed with the third, though he progressed enough to help clear a path to success for John Hall at Harpers Ferry.

Colonel Lee arrived at the armory expecting, as in the army, to have orders carried out. He found instead a collection of young men, most of them ex–farm boys, who were accustomed, except during planting and harvest seasons, to working at their own pace. Often they came to work with hard liquor in their lunchpails. When bored, they left their workbenches to gamble at cards, roughhouse, or simply idle the time away in lazy talk. Shortly after his arrival, Lee came upon two men wrestling while the rest of the shop cheered them on. Lee dismissed the miscreants on the spot and ordered the rest of the men back to their benches. Eleven refused. Lee fired them all. Detailed regulations soon flowed from his office. All men were expected to be at work from 6:00 A.M. to noon and from 1:00 P.M. to 6:00 P.M. Spirits and all extracurricular activities were forbidden. "Fighting among the workmen will not be tolerated, nor an indecent or unnecessary noise allowed in or about the shops," runs one injunction. Another reads: "Gambling of every description and the drinking of Rum, Gin, Brandy, Whiskey, or any kind of ardent spirits is prohibited in or about the public workshop." To further help tame the men, he encouraged church attendance: "Due attention is to be paid to the Sabbath and no Labor, Business, amusement, play, recreation, ... or any proceeding incompatible with the Sacred Duties of the day will be allowed"; and unofficially marriage to young ladies in Springfield was favored.

Lee knew that a number of his rambunctious young men would continue to balk at this paramilitary atmosphere. He had other inducements to keep them in harness. There was the assured pay envelope every Saturday for men who had seldom seen much cash. A constant waiting list of "some forty or fifty" young men "that are eager for a business in the Armory" helped to keep a man in line. Division of labor trained men to concentrate on a specific, often minute, task; it took the artist out of the artisan and deprived him of the mobility common among peripatetic mechanicians. "There are very few in the armory," said Lee, "that work at all branches." Lee demanded discipline. Finally, along with Whitney and other arms manufacturers in the Connecticut Valley, he endorsed the blacklist. He notified "all the masters and manufacturers to the south of Springfield not to employ each other's workmen without a recommendation from the person who employed them last." By the end of his first year as superintendent Lee had created a reasonably stable work force.

Before Lee had a chance to discipline his troops, Colonel Wadsworth convened a meeting at New Haven early in June 1815, only days after Lee's appointment to Springfield. In creating the Ordnance Department a few weeks earlier, Congress had specifically ordered it "to draw up a system of regulations ... for the uniformity of manufactures of small arms." Wadsworth passed those marching orders along to Whitney, Lee, Stubblefield (superintendent of the Harpers Ferry Armory), and other arms manufacturers at the meeting. Possibly Whitney spoke up and said that the first step toward the goal was to create special-purpose machines to produce uniform parts.

Lee, essentially more a fellow traveler than a mechanician, knew little about designing or building special-purpose machines, but he knew enough to spot men who did and to encourage their experiments at government expense. Innovations spawned by arms makers in the Connecticut Valley immediately got his attention. Shortly after the conference with Wadsworth, he had installed a trip-hammer that Asa Waters, a private contractor for the government, had adapted to welding gun barrels. Lee kept a close eye on the numerous efforts among contractors to produce gun barrels by machine. This goal raised a seemingly unsolvable problem. The cylindrical part of the barrel was relatively easy to shape, but at the breech end it flattened to an oval shape. None of at least five contractors working individually had been able to find a way to create an irregularly shaped object by machine. Asa Waters, himself an ingenious mechanician, worked more than a year trying to solve the problem. Finally, he called in Thomas Blanchard, a fellow mechanician who knew nothing about arms making but had a reputation for imaginative solutions to knotty problems. "Being told what was wanted," legend has it, Blanchard looked over Waters's machine, then "began a low monotonous whistle as was his wont through life when in deep study, and ere long suggested an additional, very simple, but wholly original cam motion, which upon being applied relieved the difficulty at once, and proved a perfect success." Lee, alerted to the achievement, immediately invited Blanchard to give a demonstration at the armory, which led to a contract to build one for the government. A year later, in February 1819, Blanchard had designed another machine that could take a rectangular piece of wood and shape it into an irregularly shaped gunstock. "Awkward looking

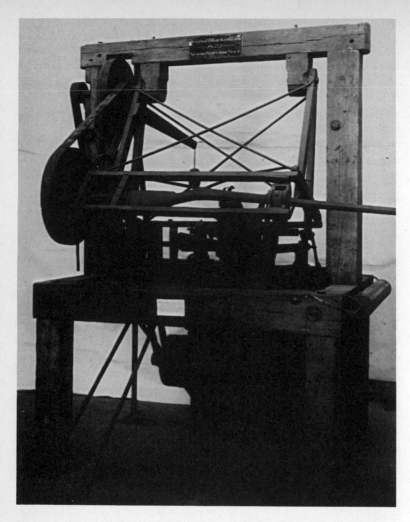

Blanchard's "lathe" to manufacture gunstocks. This awkward but revolutionary machine was built for the Springfield Armory in 1822, the prototype in what became a battery of fourteen machines that took a raw piece of lumber and shaped it into an all-but-finished gunstock. *(Smithsonian Institution)*

but amazingly efficient,'' Merritt Roe Smith writes, ''it represented one of the truly outstanding American contributions to
nineteenth-century technology.'' Four years later Blanchard,
whom Lee had given work space and free run of the armory, had
created a battery of fourteen machines that completely mecha-
nized stock-making at Springfield. ''Rarely have the contributions of one person effected so sweeping a change in such a short
period of time.''

It has been argued that Blanchard's invention was not entirely
original. Earlier Marc I. Brunel had commissioned Henry Maudslay, one of England's greatest mechanicians, to build twenty-two
special-purpose machines to produce wooden pulley blocks for the
British navy. David Hounshell holds that ''Blanchard clearly did
not draw inspiration from Brunel's machinery for his fundamental gunstock-turning lathe (because the blocks were not irregularly shaped),'' but he may have grasped from Brunel ''the
principle of sequentially arranged single-purpose machinery.''
More to the point as far as American technology is concerned,
Blanchard and fellow mechanicians saw his machines to be adaptable to a multitude of tasks—the shaping of ax handles and shoe
lasts, for example—while ''British producers failed to apply this
principle to the manufacture of other goods.'' Hounshell concludes that ''Blanchard's fundamental machine, in addition to

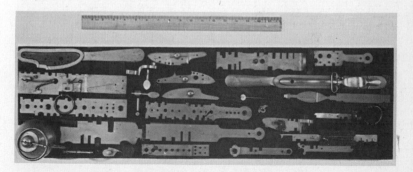

Inspection gauges used at the Springfield Armory for the United States
Model 1841 Rifle. ''This set of gauges,'' according to David Hounshell,
''was one of three different types of gauges built by the Springfield
Armory. Master gauge and work gauge sets were also constructed and
used.'' (Smithsonian Institution)

being a brilliant invention, initiated an unprecedented movement in the construction of special-purpose tools that would be used sequentially.''

Making parts by machine helped to speed production and reduce the need for skilled craftsmen, but the Ordnance Department wanted something more to assure a steady production of interchangeable parts, and at its insistence Lee inaugurated a system of gauges, which by 1819 (Smith remarks) ''had reached a point of sophistication.'' A visitor to the armory that year describes how the system worked:

> The master armorer has a set of patterns and gauges. The foremen of shops and branches and inspectors have each a set for the parts formed in their respective shops; and each workman has those that are required for the particular part at which he is at work. These are all made to correspond with the original set, and are tried by them occasionally, in order to discover any variations that may have taken place in using. They are made of hardened steel. . . . If this method is continued, and the closest attention is paid to it by the master workmen, inspectors, workmen, and superintendent, the designed object will finally be obtained.

Lee was not entirely convinced. In view of the high costs of production, he remained uncertain as he began his fifth year at the armory that the effort ''to make the muskets with that exact uniformity, that the several component parts will fit one musket as well as any other,'' was either practical or useful. Nonetheless, he went on, ''my impressions are, that this mode must be entered into and pursued until experience (which is the most sure test) proves its practicability and utility, or the reverse.'' Despite considerable achievements during eighteen years at the armory, Lee never managed to produce machine-made parts that were truly interchangeable, but he and other arms manufacturers in the Connecticut Valley blazed a trail that guided John Hall to triumph at the Harpers Ferry Armory.

15

A Visionary Theorist

Roswell Lee resisted temporary reassignment from Springfield to Harpers Ferry, but Colonel George Bomford, chief of ordnance, insisted that he come down and impose order on a chaotic situation in the southern armory. Lee arrived there, reluctantly, in mid-November 1826. He found the predicted chaos—craftsmen who gambled and drank on the job, attended cockfights and wrestling matches during working hours, and generally came and went as they pleased. He also saw at last what he had heard about secondhand, the armory's Rifle Works, a semiautonomous adjunct to the main shops, run by John Hall. Most of the men in the armory, nearly all of them southerners, disliked Hall. For one thing, he was a New Englander, also vain, egocentric, and opinionated. Finally, in their view, he was "a visionary theorist." Lee, by contrast, saw as he wandered through the works from machine to machine that Hall had realized what Lee had come to think an impossible dream—the production of weapons with interchangeable parts that came from a modest assembly line.

Hall was born in 1781, the year of the victory at Yorktown. He was a slow, thoughtful, sometimes irascible, often self-righteous, but seldom impulsive man—he waited until the age of thirty-two to marry, time enough to father seven children—who earned his living from a woodworking shop in Portland, Maine. In 1811 he designed a breech-loading rifle, a radical innovation in a day when all weapons were muzzle loaders. It took eight years to sell the army on his invention, even though extensive tests showed it

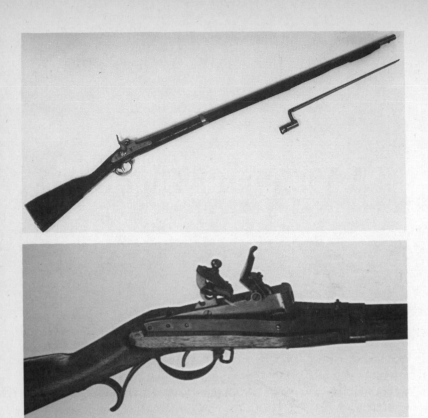

Harpers Ferry Percussion Smoothbore Musket, U.S. Model of 1842
(above). This particular weapon was found on the battlefield after the
First Battle of Bull Run, 21 July 1861. *(New-York Historical Society)*
Detail of a Hall Rifle (below). Each of the numerous metal parts of
the "lock," as it was called, were machine-made and interchangeable.
(Smithsonian Institution)

to be a lighter weapon with less recoil, easier to load, and more
accurate than the standard rifle. Along with these virtues, Hall
held out another carrot, a promise to fulfill the dream that dated
back to Eli Whitney: his rifle would be so precisely machined that
"if a thousand guns were taken apart and limbs thrown promiscuously together in one heap they may be taken promiscuously
from the heap and will all come out right." In March 1819, with
support from the supposedly reactionary secretary of war, John

C. Calhoun, Hall signed a contract to produce at Harpers Ferry one thousand breech-loading rifles, the entire project subsidized by the army, including Hall's salary of sixty dollars a month.

From the start Hall's expenses at the works shocked army inspectors. One said that the weapons "will cost the government more than 50 dollars before they are finished." True, said Hall, but his machinery when completed "will answer as well for one hundred thousand of the guns as for one thousand." Calhoun visited the Rifle Works, came away convinced Hall was no "theoretical visionary," and continued to back him. On 30 December 1822, Hall wrote to the secretary of war with understandable pride:

> I have succeeded in an object which has hitherto completely baffled (nothwithstanding the impressions to the contrary which have long prevailed) all the endeavors of those who have heretofore attempted it—I have succeeded in establishing methods for fabricating arms exactly alike, and with economy, by the hands of common workmen, and in such a manner as to ensure a perfect observance of any established model and to furnish in the arms themselves a complete test of their conformity to it.

It took four years to fill the army order, for Hall had had to start virtually from scratch. He brought in mechanicians to build machines to shape various parts of the rifle, machines so simple to operate that unskilled boys could and did handle them. He designed a system of gauges more elaborate than Lee's at Springfield, indeed "more numerous and exact . . . than had ever before been used." His greatest innovation, as David Hounshell remarks,

> became one of the fundamental principles of precision manufacture . . . the principle of fixture of design. A fixture is a device that "fixes" or secures an object in a machine tool, such as a milling machine, and holds it during the machining operation. . . . Because fabrication of Hall's rifle parts involved a number of different machine operations, . . . the inaccuracy of each fixing would be multiplied by the number of different machine operations. To rectify this problem, Hall reasoned that if the piece were located in each fixture relative to one point on the

piece, the multiplying effect could be thwarted. He called this reference point the "bearing" point and designed all fixtures for a part relative to that point. And, as Hall stressed, "this principle is applicable in all cases where uniformity is required." Indeed, this fundamental principle became universally applicable.

Roswell Lee arrived at Harpers Ferry as the army was finishing five months of tests on Hall's rifles. Its final report gave undiluted praise for the weapons. Meanwhile, the committee on military affairs in the House of Representatives, egged on by enemies Hall had created over the years, pressured the War Department to select a group of "practical judges of machinery" to investigate "every aspect of Hall's operations at Harpers Ferry," especially his "waste and extravagance of the public money." At Lee's suggestion James Carrington—"a man of strict integrity" according to Whitney, seldom given to praise—was chosen to head the committee. Its report lauded every aspect of Hall's achievement.

> By no other process known to us (and we have seen most, if not all, that are in use in the United States) could arms be made so *exactly alike,* as to interchange and require no *marks* on the different parts. . . . We would, however, further observe, that in point of *accuracy,* the quality of the work is greatly *superior* to anything, we have ever seen or expected to see, in the manufacture of small arms and cannot with any degree of propriety, be compared with work executed by the usual methods, and it fully demonstrates the *practicability* of what has been considered almost or totally *impossible* by those engaged in making arms, viz. —*of their perfect uniformity.*

Merritt Roe Smith, who has done so much to rescue Hall from oblivion, deserves to summarize Hall's achievement from the view of a modern historian:

> All told, Hall emerges as a pivotal figure in the annals of American industry. . . . No one . . . had been able to master the problem of attaining complete interchangeability in firearms. Much of the excitement generated by the special investigations of 1826 can be traced directly to Hall's success in combining men, ma-

chines, and precision-measurement methods into a *practical* system of production. . . . In this sense, Hall's work represented an important extension of the industrial revolution in America, a mechanical synthesis so different in degree as to constitute a difference in kind.

Eight years later, in 1834, Simeon North, a private contractor of Middletown, "added another dimension to the evolving pattern of precision when he adopted Hall's gauges"—they amounted to sixty-three—"and succeeded in making rifles with parts that could be exchanged with those made by Hall at Harpers Ferry." By the mid-1840s armory practice had spread to other private contractors and the mass production of muskets and rifles with interchangeable parts became "one of the great technological achievements of the modern era."

It does not diminish the work of mechanicians led by Whitney, Lee, and Hall to note, finally, the relentless support that came from the government, notably from Colonel Wadsworth and Colonel Bomford in the Ordnance Department and from John C. Calhoun. "The development of the American system of interchangeable parts manufacture," David Hounshell writes, "must be understood above all as the result of a decision made by the United States War Department . . . to have this kind of small arms, whatever the cost."

IV

The Flowering
of the North

16

Why Not the South?

Robert Beverley, a Virginian, complained in 1705 that the South did little for itself but raise tobacco. "Tho' their country be overrun with wood, yet they have all their wooden ware from England." Their hats came from England, made from furs they had sent there. They had sheep aplenty, Beverley said, but instead of converting the wool into clothing they bought what they wore from England. They imported leather goods from England and the hides from their cattle were left in the fields to "lie and rot, or are made use of only for covering dry goods in a leaky house."

The South had an abundance of natural resources, yet little seemed to change in the ensuing centuries except the source of their imported goods. Sometime after the Civil War a southern newspaper editor described a funeral he had attended in rural Georgia. "The grave was dug through solid marble, but the marble headstone came from Vermont," he wrote.

It was in a pine wilderness but the pine coffin came from Cincinnati. An iron mountain overshadowed it but the coffin nails and the screws and the shovel came from Pittsburgh. With hard wood and metal abounding, the corpse was hauled on a wagon from South Bend, Indiana. A hickory grove grew nearby, but the pick and shovel handles came from New York. The cotton shirt on the dead man came from Cincinnati, the coat and breeches from Chicago, the shoes from Boston; the folded hands were encased in white gloves from New York, and round the poor neck which had worn all its living days the bondage of lost opportunity, was twisted a cheap cravat from Philadelphia.

That country, so rich in underdeveloped resources, furnished nothing for the funeral except the corpse and the hole in the ground and would probably have imported both of those if it could have done so. And as the poor fellow was lowered to his rest, on coffin bands from Lowell, he carried nothing into the next world as a reminder of his home in this, save the halted blood stream in his veins, the chilled marrow in his bones, and the echo of the dull clods that fell on his coffin lid.

Why, the editor wondered, had his beloved South remained through the centuries a colony, once England's, now the North's?

Actually, the South depended less on manufactured goods drawn from the outside than the editor assumed. On the eve of the Civil War its imports totaled no more than 15 percent of its economy, and of that 15 percent the South took no more than 8 percent from the North. Nor did it always resist technological innovations. John C. Calhoun early on urged the South to adopt them. He spearheaded the successful campaign in Congress for a protective tariff that allowed Lowell's mill in Waltham to compete profitably with the British. In 1818, Nathan Appleton recalled forty years later, Calhoun visited Waltham "with apparent satisfaction of having himself contributed to its success." Calhoun went on, as secretary of war, to encourage John Hall's costly plan to produce muskets and rifles with interchangeable parts at Harpers Ferry, but his views on the virtues of technology and industry for the South swiftly changed. "It is lamentable," Appleton went on, "to think that in 1832, under the alluring vision of a separate confederacy, he should have become the active enemy of the manufacture which was doing so much for the interest of the planters, and that the influence of his name has continued to keep them in error."

An innovative spirit throve in the South from the earliest days. At a time that England had none, Virginia erected the first sawmill in the New World shortly after the settlement of Jamestown, and one of its first shipments to England was a cargo of mechanically sawed lumber. A northerner invented the cotton gin, but the South quickly violated the patent and made cotton its major

cash crop. Early in the nineteenth century it built mills that
converted raw cotton into thread or yarn and later, in more complex mills, into cloth. Grist mills scattered along scores of streams
and rivers produced flour; an English traveler in the 1840s reported on one mill that ground a million bushels of wheat a year.
A metal shop in Georgia owned by Robert Findlay in the 1850s
produced gears for cotton gins, castings for plows, steam engines,
and boilers, screws to press cotton into bales, cane crushers, syrup
boilers, pipes, pumps, circular saws, and railroad castings. On the
eve of the Civil War, the Tredgar Iron Works in Richmond
grossed more than $1 million a year. In 1853, the year of the
Crystal Palace Exhibition held in New York City, Louisville had
its own exhibition, displaying buggies, stoves, plows, and a variety of other manufactured goods built by local craftsmen.

Most of these enterprises made money for their owners, yet the
South only dabbled in a mechanical world the North was embrac-
ing. In 1860 Lowell, Massachusetts, alone had more spindles turning out cloth than the entire South, according to one scholar.
(This figure, as so much in the story of the nuts and bolts of the
American past, could be suspect, for another scholar holds that
before the Civil War "the slave states produced between 20 and
25 percent of the national manufacture of cotton textiles.") All
the machinery in Robert Findlay's metal works in Georgia had
been designed and built in Philadelphia shops. The South refused
to leap with the North into the industrial age, despite its abundance of natural resources—large deposits of accessible coal, iron
ore, and limestone; vast reserves of timber, innumerable streams
to turn its water wheels. But technological and industrial advancement, Neil Harris has said, "seemed least appreciated" in
the South; "there the smallest proportion of the population was
exposed to revolutionary physical improvement or accustomed to
tinkering with machinery. Moreover, the mechanic himself, along
with the laborer, enjoyed less status there than elsewhere because
of the taint placed by chattel slavery upon manual labor."

The mercantile class and owners of large plantations had sufficient capital to develop the South's resources. No shortage of
labor prevailed. The huge population of black slaves, often idle in
the winter, was there to lease out to shops and mills if any existed
in the neighborhood; in *Uncle Tom's Cabin,* Harriet Beecher

Stowe has George, a slave, spending the slack season working in a Kentucky shoe factory. Whites resisted working alongside blacks, but their resentment could be overcome. The owner of the Tredgar Iron Works "broke the opposition of his free workers and introduced large numbers of Negro slaves into skilled rolling mill positions."

The South, for the most part, turned its back on these assets and continued to plow its money into land. "The agrarian ideal that permeated all classes of southern society undoubtedly hampered the growth" of industry, Clement Eaton has said. A northern visitor, listening to barroom talk one evening, learned that an entrepreneur was something less than a man because he "lived by being complaisant and civil to all, and when he ceased to be so he starved, while the farmer who was not dependent on others for his living could be as aristocratic and haughty as he pleased with perfect impunity." A number of publicists, notably William Gregg, who owned profitable textile mills in South Carolina, did all they could to promote industry but with little success. (Gregg, incidentally, refused to hire any nonwhite workers for his mills.) The landed gentry dominated politics and refused to let legislatures give comfort and aid to industry. They had cash to spare, particularly in the 1850s, when agricultural prices were high, but saw no reason to invest it in risky ventures that they held in disdain and knew little about. Also, as Diane Lindstrom has noted, "The tremendous improvement in their terms of trade, that is, their ability to purchase more imports with the same quantity of exports, eroded whatever incentive existed to develop manufacturing between 1815 and 1860."

Even had the incentive existed, the chance that industry could flourish in the South was slim. It lacked a pool of trained, disciplined workers. The southerner who invented "the Virginia Reaper" went north to manufacture it. The chief engineer of a Virginia railroad reported in 1835 that the need for skilled labor "has prevented our perfecting and extending the arrangements which were contemplated a year ago for the construction of locomotives, and the wheels and axles for cars." When the Tredgar Iron Works set out to employ slaves, it had to import mechanicians from the North to teach them the skills of puddling and rolling iron. The South needed a large transfusion of fresh blood to undermine the agrarian ideal, but it received only a trickle. While thousands of immigrants poured into northern towns and cities, many bringing the latest technological innovations from Europe, only a few drifted into the South. Robert Findlay came from Scotland and learned the machinist's trade in Philadelphia before he set up his metal shop in Georgia. Daniel Pratt, a New England carpenter, went to Alabama and created an industrial village, where he manufactured cotton gins and also built a cotton mill, a gristmill, and a foundry, as well as machine, blacksmith, carpenter, and various other shops, all of which supported a community of eight hundred people. But these gentlemen were exceptional. Most who came down from the North found the South inhospitable to dirty-fingernail people. The typical mechanician too often found himself without work and to survive either headed back North or out West. Also too often even those who prospered failed to stay. J. Edgar Thomson from Pennsylvania spent fifteen years building the Georgia Railroad but made his mark in life as the first great president of the Pennsylvania Railroad.

The South in time might have overcome its resistance to the new technology and the people who were spreading it about the nation, but it still had to face a deficiency that was to plague it throughout the nineteenth century—the lack of what Diane Lindstrom calls "growth poles"—namely, towns and cities that, as she defines the term, "enjoyed a well-developed business system, relatively underemployed labor and capital, a high level of artisan and mechanical skills developed indigenously but cross-fertilized by a stream of skilled British machinists and textile workers, and populous and prosperous hinterlands well-served by inexpensive water transport." None of the four southern towns that were receptive to technological innovations and that contained an embryonic industrialism—Richmond, New Orleans, Charleston, and Louisville—satisfied all of these requirements, particularly the last one, an accessible and prosperous hinterland that could buy what a "growth pole" produced. Louisville's appeal in 1856, through its Mechanics' Institute, brings tears to the eyes of anyone who wished that part of the country well: "Louisville is at this time more fully embued with the spirit of fostering and encouraging MANUFACTURES AND MECHANICS than ever before.

The action of the Chamber of Commerce to make Louisville the Manchester of the West, has met with the sanction and hearty cooperation of all classes of our citizens. The SPIRIT OF PROGRESS is aroused. The Manufacturer's, the Merchant's, the Mechanic's and the Artist's motto now is—ONWARD.''

The well-to-do still preferred goods made in England; the blacks had no purchasing power and the poor whites only enough to satisfy basic needs. William Gregg spotted this deficiency. ''An exclusively agricultural people in the present age of the world will always be poor,'' he wrote. ''They want a home market. They want cities and towns. They want diversity of employment.''

The workshops clustered within a ''growth pole'' like Philadelphia were invariably small, employing from five to fifty workmen, yet, small as they were, incredibly versatile. Oliver Evans's Mars Works produced a variety of items tailored to customers' needs—steam engines; machinery for sawmills, forges, rolling and slitting mills, sugar mills, apple mills, bark mills; pans for sugar boilers and soap boilers; screws of various sizes for cotton presses and tobacco presses; and all kinds of wheels, gears, and machinery for cotton and wool mills. Much of what he produced ended up in the South.

Flexibility as well as variety marked the shop's production. What Jane Jacobs says of workshops in modern cities holds for Philadelphia's in Evans's day—constant improvisations and interactions occurred as an everyday matter. She draws on Charles F. Sabel's account of small firms in industrial cities of Italy today to make her point. '' A small shop producing tractor transmissions for a large manufacturer modifies the design of the transmission to suit the need of a small manufacturer of high-quality seeders. In another little shop''—here she quotes from Sabel—''a conventional automatic packing machine is redesigned to fit the available space in a particular assembly line. A machine that injects one type of plastic into molds is modified to inject a cheaper plastic. A membrane pump used in automobiles is modified to suit agricultural machinery. A standard loom or cloth-cutting machine is adjusted to work efficiently with particularly fine threads.'' (Incidentally, neither Jacobs nor Sabel needed to have gone to Europe for illustrations. An early coworker of Henry Ford's once recalled the shop where they both broke in as mechanicians. ''It

was a great old shop,'' he said. ''They manufactured everything
in the line of brass and iron—globe and gate valves, gongs, steam-
whistles, fire hydrants, and valves for water pipes. . . . They made
so many different articles that they had to have all kinds of
machines, large and small lathes and drill presses. . . . They had
more machines than workmen in that shop.'')

Evans competed with workshops in the neighborhood for business, yet a steady exchange of ideas passed among them. He could
walk across the street or down the block to air a problem or suggest a solution to a colleague's problem. All of these shops depended on one another for more than ideas. The son of one of
Evans's contemporaries recalls their close-knit relationship. Coleman Sellers produced a line of products nearly as diversified as
Evans's. One season he might be building fire engines, another
time the machinery for a paper mill. The housings for the paper
mill, his son recalls, ''were cast at Park's foundry, Kensington;
most of the gearing at Wilberger's foundry, S.E. corner of Market
and 16th Street. The housings were chipped and filed at Boyle's
shop where most of the smith work was done; there were no gear
planers at that time. The gear wheels were bored and fitted at our
little shop at Cardington and either Jesse Hayes, William Lund-
gren or Caulkins were the millwrights who made the vats and put
up and started the machinery.''

The constant give and take of ideas and advice among these
mechanicians of the past—and among those of today, it should be
noted—led to an elastic and constantly changing use of the latest
technology. The innovative capacity of these small firms, Sabel
remarks in another context, was spurred on by their ''close relations with . . . similarly innovative firms in the same and adjacent
sectors; and above all on the close collaboration of workers with
different expertise.'' All these factors were lacking among southern mechanicians in the nineteenth century.

17

Machines in the Garden

Forget for the moment the noisy clanging within the Sellers
and Evans machine shops in Philadelphia; forget the huge brick
buildings at Waltham and Lowell where thousands of men and
women toiled. Anthony Wallace describes a more typical scene in
early industrial America. Down to the 1840s "a harmony of rural
and industrial interests and lifeways, nearly became a reality,"
he writes. Most of the textile mills and machine shops stood in a
bucolic setting, scattered along creeks and small rivers, "and each
employed dozens rather than hundreds of workers." The workers'
"way of life and their attitudes and values were still close to
those of their neighbors, who continued to be full-time farmers."
Their machines were powered by water, "a non-polluting, self-
renewing source of energy." Indeed, Wallace concludes, "The
machine was in the garden, to be sure, but it was a machine that
had grown almost organically in its niche like a mutant flower
that was finding a congenial place among the rocks, displacing no
one else and in fact contributing to the welfare of the whole."

Wallace speaks of the scene along creeks in a rural section south
of Philadelphia, but his picture holds for other parts of the country. Zachariah Allen, writing of New England in 1829, said that
manufacturing there is "carried on in little hamlets, which often
appear to spring up in the bosom of some forest, gathering around
the water fall which serves to turn the mill wheel." Later, a
historian surveying five hundred New England townships said it
would be hard in 1840 "to find 50 . . . which did not have at least

124

one manufacturing village clustered around a cotton or woolen mill, an iron furnace, or a carriage shop," all driven, of course, by waterpower.

The owners were a diverse lot who seldom had more in common than a previous experience in business and more often than not at least one bankruptcy behind them. Entrance into the textile business cost little. A defunct gristmill or fulling mill could be rented and renovated for a small sum, and friends who still had faith often put up the cash to buy machinery. These entrepreneurs, usually fairly young, were shrewd, bright, and hard-working. As on-the-scene owners, they lurked constantly about the premises learning all they could about the esoteric world in which they had invested. They read whatever technical literature they could find; essays like "Observations for Carding, Roving, Drawing, and Spinning, all Kinds of Cotton Twist" became secular bibles.

None of these small mills sought to compete with the behemoths at Waltham and Lowell, which daily turned out thousands of yards of cheap but sturdy cloth for a mass market. The small mill owners produced what would sell in local markets. Virginia mills concentrated on yarn because farm wives, with looms in their homes, wanted to make their own cloth. Mill owners close to a city listened to what retailers told them customers wanted, especially in the way of fancy goods. Intense competition and volatile markets forced these small mill owners to be more flexible than the giants. "In the 1820s capital- and power-starved," small mills in southern New England "adopted the Taunton speeder and its close relatives, the plate and eclipse speeders, both because these were far cheaper to buy and operate than the Waltham double speeder favored at Lowell and because they used less power," J. W. Lozier has written. "In the 1830s and 1840s, while southern [New England] mills shifted to ring and Danforth frames, which could be driven at higher speeds while using less power, northern mills stayed with their slow, heavy throstle frames." Lozier's remarks sound like double-talk to a layman, but they make the point that those who owned the machines in the garden were as innovative and certainly more flexible than the giants, whose size impeded basic changes. They also suggest that machinery alone led to more efficient production. On this point Betsy Bahr argues from her study of a single small mill that the introduction of

machinery "did not necessarily lead to a significant gain in the mill's overall volume of production." The machinery, though not "economically justified," did have a virtue—it displaced skilled hands with unskilled ones who could be made competent at their posts after a few days of training.

Typical mill workers contributed little or nothing to technological innovation. Most regarded the mills as way stations where they paused only long enough to earn a stake that would lead to more appealing work. One man, who had been a stonemason in Ireland, departed after less than a year but left a wife and offspring behind in the mill until he had reestablished himself. Another after three years returned to the land as a farmer. The tedium of mill work offered little chance to learn a skill that could lead to advancement, which helps to account for the enormous turnover in the labor force. Statistics vary from mill to mill, region to region, but an annual turnover of 50 percent was not uncommon. Wallace estimates that nearly the entire work force in the Rockdale district renewed itself every decade.

The workday began at 5:00 A.M. and ended at 7:00 P.M., a routine broken by a half-hour pause for breakfast and forty-five minutes for the midday meal. Everyone worked a six-day week. The children often fell asleep at their posts. All windows had to be kept shut to prevent the blowing about of cotton fibers, often soaked with oil from the machines. The rooms were comfortable in winter but stifling in summer. The women often worked nude to the waist. The thick air filled with cotton particles often left new hands feeling "a little nausea at first and their appetite for food is lessened." No one was allowed to smoke.

Many historians hold that the mill routine created a new type of worker—disciplined, abstemious, and reliable. William Sisson, from a study of two small mills, argues against this view. Workers "did not find it necessary to restructure their work habits and

In the top frame of this montage, a small locomotive can be spotted, but in the central bucolic scene no sign of the railroad, which brought raw materials into Shaftsbury, Vermont, and carried out finished products, mars the setting. The advertisement admits that the chisels and steel squares were made in a factory but one set in the country where old-fashioned rural values prevailed. *(New-York Historical Society)* values severely. They continued to act by the same ideas of time, discipline, and temperance that they held before they entered the mills.'' About the sharpest break with past customs was an erosion of the patriarchal system within families. Women, now that they contributed substantially in cash returns to the family welfare, in take-home pay, acquired more say in family decisions. They usually managed the money.

Posted rules and regulations made day-to-day life sound more harsh than it was. Lateness was forbidden, no one was allowed to leave a station without permission, and so forth; yet workers often left their machines on a warm day to fish, despite posted rules and regulations that forbade such action. It was not unusual for a man to show up drunk or so badly hung over that he was all but useless for the day. No swift jump occurred from rural to industrial customs. The factory system settled unobtrusively into a rural setting, and every job in every mill had built-in if unscheduled holidays. Lozier notes that in a ten-month period one man's station at a Delaware cotton mill ''was shut down forty-one days for mechanical malfunctions, waterwheel repairs, ice, floods, and shortages of slubbing [thick nubs of yarn] and oil.'' Another owner had to stop operations for five weeks while he had his waterwheel rebuilt, giving employees a lengthy but payless vacation.

Mill families, according to Wallace, lived fairly well. Men earned about $2.50 a week, women about half that, and children about half their mother's salary. Rent was about 20 percent of a man's salary, roughly $2 a month. Food costs were in the region of fifteen cents a day if a family grew its own vegetables and had a cow that gave milk. Compared with city laborers, who were ''truly poor, skating on the edge of starvation,'' according to Billy G. Smith, they lived very well. ''The general picture which emerges is hardly one of increasing exploitation or dire poverty among these early manufacturing operatives,'' Donald R. Adams, Jr., concludes. ''Workers were able to accumulate at a surprisingly rapid pace. On the average workers could save a year's earnings in just over six years without interest''—the Du Pont company paid employees 6 percent interest on savings banked with the owner—''and in less time with interest.... Household units might accumulate at even more rapid rates if other members contributed earnings.'' Wallace tells of one man who singlehand-edly saved enough to take his family of four to Ohio (which in 1836 cost $21 by canal and rail) and to buy 120 acres of farmland for $150 and a pig and two sows for $11.50, and had enough left over to pay workmen to build a log house.

18

The Machine Shop

Every stretch of waterway that hosted a line of mills along the East Coast had, within a day's travel, a machine shop ready to repair broken equipment that a local superintendent found beyond his abilities to put back on line. The Hodgson brothers served numerous mills along the Brandywine, south of Philadelphia. Shops in Pawtucket and Providence cared for those of Rhode Island. West Point had a machine shop that nurtured upstate New York and Western Connecticut. Mechanicians' shops dotted the landscape of the Connecticut River from the Vermont border to the river's outlet at Old Saybrook. Notable shops existed in Paterson, New Jersey; Saco, Maine; and Manchester, New Hampshire. When a serious breakdown occurred at any mill, the nearest shop sent along a mechanician. While repairing the machine, he might suggest an innovation that he or someone in his shop had come up with that could cut the owner's cost and speed up production. The mechanician, more than anyone else who infiltrated the small mills, kept the owners alive to the new world that was coming into being.

To say so is not to denigrate the role of the entrepreneur in advancing technological innovation. He, ultimately, had to decide when to innovate and when not to. He had to be, as David Noble calls a latter-day entrepreneur, "a born promoter"; also someone who had the "common sense, drive, salesmanship, and confidence" to keep his enterprise profitable. Technological and social revolutions, Noble goes on, have one thing in common: "They do not simply happen but must be made to happen. The enthusiasms of the people who drive them must overcome the resistance of reality, that is, of other people's reality.'' That enthusiasm, an often overlooked aspect of America's technological success in the nineteenth century, coupled with that of the mechanician, calls for emphasis.

By the 1830s or 1840s the machine shops had acquired a fixed form and a small flotilla of machines—the treadle-powered lathe young George Sellers had known was now supplemented by steam- or water-driven lathes, drill presses, planers, drop hammers, screw augers, and the like—that were to stay little changed until the Civil War and in many shops long afterward. Otherwise, much remained as it had in Sellers's youth. The shops were still shops, seldom employing more than half a dozen men (Lowell's, one of the largest, had two dozen machines but no more than seven mechanicians to operate them), and the mechanicians were still lords of the working world, some arriving at their posts wearing top hats and flowing cravats. No one worried about safety measures—goggles to shield the eyes, hard-tipped shoes to protect toes from dropped objects—any more than they had in young Sellers's day, when one morning he spied a worker ''with his head close to the shaft he was turning with his arm and hand frantically trying to reach the belt shifting lever. I jumped quickly to the lever, shifted the belt, backed the lathe by hand and when I got his long ended black cravat which had caught in the carrying dog unwound he fell to the floor limp and insensible and it was some time before we brought him to.''

Most of the machines in any shop had been modified from stolen British prototypes. An exception was the milling machine, which historians agree Simeon North of Connecticut invented about 1816. ''Until then,'' says David Hounshell, ''the principal way to remove metal in shaping it was by filing, using either hand or rotary files. As developed, milling technology allowed flat or curved surfaces to be cut by simply passing a revolving hardened steel cutter over the iron, resulting in uniform and quickly executed parts.'' In creating the machine, ''North helped set American manufacturing on the road toward mass production.'' Another exception was Thomas Blanchard's lathe for making gunstocks, which inaugurated in American shops tracer control.

"Whenever a particular contour was required in a machine operation," Noble explains, "an appropriate template was traced by a stylus and the programmed information was automatically conveyed to a cutting tool, which worked much like machines used to duplicate keys, where the original serves as the template for the duplicate. To change the contour it was only necessary to change the template."

Gear-cutting machines developed first in Europe, but special mention should go to one designed by Joseph Saxton of Philadelphia, who, said a contemporary, "is justly celebrated for his excessively acute feeling of the nature and value of accuracy in mechanism, and who is reported not to be excelled by man in Europe or America for exquisite nicety of workmanship." The gear-cutting machines opened a variety of doors for mechanicians. "The availability of accurate gearing in quantity," Robert Woodbury has said, "made possible the sewing machine, the rotary printing press, the automobile, and many other familiar engines."

"A machine then," remarks a modern visitor to a shop filled with machines built in the nineteenth century, "was clearly a machine: it was oily, smelly, dirty; it had wheels that turned, belts that flapped; it rattled, shook, roared, screeched, and squeaked. Smoke came out. Today, activity is signalled by the silent blinking of light-emitting diodes." To an outlander the old machines are only animated clumps of metal. To the mechanicians bent over them, each has "its own unique personality," as Robert Pirsig put it, his thoughts induced by repairing a motorcycle:

> This personality constantly changes, usually for the worse, but sometimes surprisingly for the better, and it is this personality that is the real object of motorcycle maintenance. The new ones start out as good-looking strangers and, depending on how they are treated, degenerate rapidly into bad-acting grouches or even cripples, or else turn into healthy, good-natured long-lasting friends. This one, despite the murderous treatment it got at the hand of those alleged mechanics seems to have recovered and has been requiring fewer and fewer repairs as time goes on.

The mechanician continued, as in Sellers's youth, to be a generalist rather than a specialist, able to operate comfortably any machine in the shop. He was still an artist who sought constantly to satisfy his "own exacting standards of workmanship." He also continued to be a peripatetic, roaming from company to company soaking up new experience. Henry M. Leland grew up on a New England farm, began as a mechanician in a loom factory, moved on to the armory at Springfield, to the Colt factory in Hartford, to a wrench company in Worcester, then to Providence and Brown & Sharpe, where he headed the sewing-machine department. At every stop along the way, "My vision of the possibilities of manufacturing broadened," he said in old age. "I realized that manufacturing was an art and I resolved to devote my best endeavors and my utmost ability to the Art of Manufacturing." He ended up in Detroit where he designed and built the first engines for the Cadillac automobile.

For all they had in common, Leland and George Escol Sellers would have had difficulty communicating about their work, and the difficulty would not have been because of age differences. Leland' roots in the past went as deep as Sellers's; he had worked with men who had known Whitney and Roswell Lee. His experience at the Springfield armory and other New England shops had exposed him to precise machining that produced interchangeable parts. Sellers worked with a wooden rule that gave him tolerances no better than a tenth of an inch. This degree of accuracy was all right "if one is making a wheel barrow with its one wheel," said Leland; "it is of no importance whatever whether the wheels are the same size on each wheelbarrow or whether they vary half an inch in diameter. On the other hand, if for instance one is making a wristpin to go into a piston in an internal combustion engine, he must make it to a close fit or it will be an absolute failure. So in making the piston-pin we set the limit as close as one-tenth of a thousandth of an inch in diameter." Leland represented a new breed of mechanician that came to maturity as the Civil War approached.

V

The Multiplication
of Things Unfamiliar

19

The Southerner in the North

Cyrus Hall McCormick was born in 1809 in a rural southern county that during the Civil War became part of West Virginia. His father, Robert, a prosperous farmer and also a self-taught mechanician, was happy to leave field work to his eighteen slaves while he tinkered at the forge in his blacksmith shop. At some point he became obsessed with the notion of reaping grain with horse-drawn machines. The process of bringing in a harvest had changed little since the Middle Ages. In America, the sickle had been supplanted by the scythe and then the cradle, but it still took scores of men, women, and children to bring in the crops during the few days in which the ripe grain had to be harvested or lost. Robert McCormick, a thrifty man and perhaps inspired by a fellow Presbyterian, Dr. Benjamin Rush of Philadelphia, had dared to alter one deep-rooted custom of the harvest season. Rush, beginning in 1782 and annually until he died, deplored "the expense of drenching reapers for two or three weeks with spirits." It cost as much as one-fourth of a farmer's profits to keep the crew in rum and whiskey. Slake everyone's thirst with water, buttermilk, cider, or table beer, said Rush, and McCormick was among the few farmers in hard-drinking Virginia who had accepted that advice.

The dream of shifting from the scythe to a horse-drawn reaper that would do what large, often drunken, crews had done for centuries would, if realized, revolutionize commercial farming. No one knows how far old man McCormick had progressed toward his goal before son Cyrus, then twenty-two stepped in. Another son, Leander, years later said that Cyrus had usurped credit due to the father. It may have been Robert who made the first great leap to the notion that a reaping machine that tried to duplicate the swinging motion of a man's sickle or scythe would never succeed. The machine, as Roger Burlingame reconstructs the thoughts of father and son, "must have shafts to one side so that the horse could not trample the grain." More important, it "must cut with moving knives set on straight bars at right angles to the line on which the horses moved. They must be made to come together with a kind of scissors motion." Finally, "there must be a device to straighten out the grain so that the knives could get at it . . . a large revolving reel which straightened out the windblown or fallen grain ahead of the cutters."

After a decade of experimentation, during which Cyrus took out a patent, the McCormicks felt ready in 1841 to market their machine. They built two that year, seven the next, twenty-nine the following, and fifty in 1844. During these years Cyrus traveled the road, constantly exhibiting the family creation. He sought, as Harold C. Livesay notes, to market one of the most

An early Virginia reaper. *(New York Public Library)*

Cyrus H. McCormick. *(New York Public Library)*

complex, delicate machines ever offered to a public that knew little about machinery. Naturally, the field trials revealed flaws—hidden stones could nick, even shatter, the knife blades; the noise made horses skittish; violent vibrations on bumpy, hilly fields often shook wheat kernels from the stalks—but McCormick refused to be dispirited. He saw a market out there among normally tradition-bound farmers and determined to capture it before his patent ended and before Obed Hussey, who had built a better machine, beat him out. He lacked capital to expand but devised a clever way to slide around that obstacle. He sold territorial rights to machine shops through New York, Pennsylvania, and Ohio and as far west as Chicago to build his machines. Within a year production jumped to 190 machines. Their quality, however, varied from shop to shop and many, if not the majority, were defective. McCormick saw that he must centralize production to assure quality, and he soon went into partnership with a mechanician in

Chicago, who was turning out the best work. The shop produced five hundred machines for the 1848 harvest. McCormick soon broke with his partner—he had no desire to share the fortune that loomed ahead—and brought in his brothers, William to handle business affairs and Leander to run day-to-day operations. In 1851 the expanded shop turned out a thousand machines.

McCormick gave only a sidelong glance at the way Leander ran the plant, but no one doubted who owned the business, who made final decisions. His ingenuity as an entrepreneur led the company to dominate an industry that by the eve of the Civil War had at least one hundred competitors. A reaper in the 1850s cost about $130. Few farmers could afford to pay cash for it. McCormick invented the installment plan—buy now, pay later—knowing that his land-bound customers were good risks. He invented the franchise system to assure farmers they had a knowledgeable agent on hand when the reaper needed repairs. He bought up rights to innovations when they came on the market and forced Leander to incorporate them into each season's new models. Few of these changes were basic, but, as has been observed, "the alterations made annually are evidence of the never-ending experimental work in progress and the presence of competition." Cyrus saw a huge market out there waiting to buy his reapers and kept prodding Leander to turn out more of them. Why, when production seldom passed five thousand, could not the plant build twenty thousand, thirty thousand, even forty thousand, machines a year? Leander had a sound answer : the constant model changes made it impossible to mass-produce a machine crafted and assembled by skilled mechanicians.

Cyrus the optimistic salesman—"don't be scared man!" he told his brother—and Leander the pessimistic mechanician argued bitterly through the 1850s, but their common southern background bound them even tighter than blood ties. Their accents allowed neither to hide his origins, but as the antislavery movement intensified after Harriet Beecher Stowe published *Uncle Tom's Cabin* in 1852 they tried to soft-pedal it. McCormick's name for his machine—the Virginia Reaper—vanished from advertisements before the decade ended. He let the public know that he gave generously to the Presbyterian church in Chicago but not that he continued to own two slaves back home, whom he leased out annually for $110. When the Civil War came, Cyrus could not face the debacle and abandoned America to live in Europe, ostensibly to promote production of his machines there. Leander complained that he had fled ''away from this land of blood and death —where we are trodden by abolitionists in the *North*—without liberty of speech—and with utter ruin in the South.''

Leander was an able man imprisoned in his past, and in a practical way the southern upbringing affected him in day-to-day affairs more deeply than Cyrus. He created and ran the Chicago plant much the way he and his father had constructed prototype reapers in the family blacksmith shop. Indeed, David Hounshell has remarked in a perceptive essay, he operated ''the reaper works as though it were a large country blacksmith shop.'' His abhorrence of the antislavery sentiment developing in New England must have helped to isolate him from the new technology arising there. ''Rarely,'' Hounshell goes on, ''did he draw upon the stock of knowledge about large-scale manufacture that had developed in New England, and even more rarely did he use the special tools devised there for the New England concepts of specialization in machine tools.'' The bulk of the employees were skilled mechanicians accustomed only to general-purpose rather than special-purpose machines, and despite McCormick advertisements that all parts of the reaper were machine-made and interchangeable the fact was that ''a significant amount of handwork went into completing the Virginia reapers.''

The Chicago fire of 1871 destroyed the McCormick plant and gave them the chance to start afresh. They failed to use it, and not until 1880, when Cyrus finally abandoned Leander, and the road to North and South reunion had been completed, did the company begin to notice what had been going on in New England. It hired a man who ''has been with the Colt Firearms Co., the Connecticut Firearms Co., the Wilson Sewing Machine Co., and many other concerns.'' In 1879 the McCormick works had turned out 18,760 machines; a year later it had produced 30,793. And a year later, though the new man had left, he had passed on what New England had to teach, and Cyrus's son had raised production to nearly 49,000 machines a year.

20

Another
Song-and-Dance Man

It can be argued that of all the song-and-dance men who pervade the early history of American technology Samuel Colt deserves to head the list. He was a showman before he could spell the word. As a youngster fascinated with explosives he set off an underwater mine in a pond that astonished local citizens and probably paralyzed his parents. Later, when down on his uppers, he traveled the country as "Dr. Coult," giving "scientific" lectures that centered on the marvelous properties of "laughing gas" (nitrous oxide). When he hired Eli Whitney, Jr., to fill an army contract for one thousand revolvers, he made sure Whitney got no credit for the work; he had every weapon stamped "Address Samuel Colt, New York." (Colt knew that New York had more prestige everywhere than the village of Whitneyville, Connecticut. Colt's flight back to the backwaters of New England may have been hastened by Henry Jarvis Raymond, an early editor of the *New York Times,* whom Richard Kluger describes as a small, black-bearded fellow with a face "no bigger than a snuff-box." Raymond, Kluger goes on—without noting that the editor has his own definition of "all the news that's fit to print"—"distinguished himself in reporting on the trial of one John Colt, brother of the six-shooter inventor, who had been charged with murdering a printer over a trifling debt and, after his conviction, cheated the gallows by taking his own life with a knife three hours before the scheduled execution.'') In 1850, when rich and famous, Colt helped a friend get elected governor of Connecticut by escorting him home before he got drunk from banquets on the campaign trail; he rewarded Colt after the election with the title of colonel. Colt, knowing the British affection for titles, introduced himself everywhere as "Colonel Colt" when he went to England to display his revolvers at the Crystal Palace Exhibition in 1851. Even Colonel Sanders would have envied his showmanship.

Colt was what today would be called a problem child, always in trouble, and with the hope that the sea would tame a fractious nature his father apprenticed the sixteen-year-old lad to a year before the mast on a ship sailing from Boston to Calcutta. During the tedious voyage—Colt forever after hated the sea—he carved from wood his first revolver. He never made clear where he got the idea—possibly from the helmsman spinning the ship's wheel, possibly from watching the revolving grinder of a pepper box. He returned home in 1831 with his model and sent a plan of it to the patent office. A mechanician made for him—Colt never numbered

Samuel Colt. *(New York Public Library)*

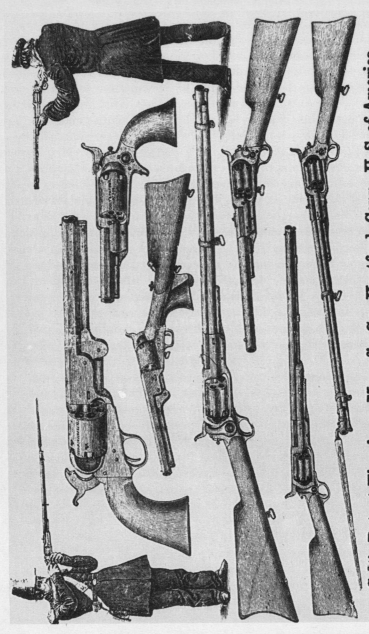

among the dirty-fingernail people—"working models of both a revolver pistol and revolver rifle," and in 1836 Colt received a patent. Soon after, the song-and-dance man persuaded a group of businessmen to back the inventor, and together they formed the Patent Arms Company in Paterson, New Jersey. The timing could not have been worse. The army had rejected his weapons, and the nation was deep in the misnamed Panic of 1837—really a depression that lasted for five years. Despite Colt's assiduous courtship in Washington of congressmen and officials in the War Department, he could not get a government order. The company produced some five thousand weapons and sold them all but not at a profit large enough to outlast the depression. It went bankrupt in 1842, and "Dr. Coult" returned to the road in order to survive.

Colt was still in his twenties—twenty-eight, to be exact—when the company folded. He had learned three lessons, at least, from failure. First, if he got a second chance the new company would be his in name and fact. He would own, control, and run it. Second, he now knew there was a market out there for the revolver that he, an easterner, had never suspected. As early as 1839 enough of his weapons had found their way into Texas that they became known as the "Texas pistol" or "Texas arm." Early in the 1840s Captain Samuel H. Walker of the Texas Rangers had come to New York to resupply his men with revolvers. He met Colt and told him that the original model had several flaws, that "it was too light; that, as it was in three pieces while being loaded, a mounted man was very liable to lose a part; that a trigger guard was necessary; that the strength and weight of the pistol should be such as to render it serviceable as a club when empty." (Colt generously named the new, 1842 model the "Walker pistol.") Finally, he learned much about manufacturing. The Paterson company had hired an ingenious mechanician named William Ball to run the shop, "one of the most prolific inventors and designers of machinery [Colt] ever knew." The lesson Ball taught —let machines do the work of men—stayed with Colt.

Four years of desuetude ended in 1846 when President Polk led America into a war with Mexico. Texas Rangers attached to

A page from Colt's catalog. *(New York Public Library)*

General Zachary Taylor's army pressured the old man for an issue of revolvers—or "six-shooters," as they had come to be called in the Southwest. Taylor's pressure on the War Department led to a contract with Colt for one thousand weapons at a total cost of $28,000. The contract caught "Dr. Coult," for once, flat-footed. He lacked a single revolver. "I advertised in the newspapers for a specimen of my own arm," he said later, "as I had given my samples and models all away to friends, but I did not find one at the time." He whittled out a new model and then made an astute arrangement with Eli Whitney, Jr. Whitney would design and build the machinery to produce the weapons, but Colt would own it. Colt lost $3,000 on the contract but ended up with Whitney's machines, which he moved to an abandoned textile factory in Hartford, Connecticut.

Colt ran his new company—named, naturally, after its founder —into a fortune. Quietly, early on, without fanfare, and with the bait of any "such compensation as you think fair and reasonable," he drew into the business to run day-to-day operations a mechanician well known through the Connecticut Valley, Elisha K. Root. Root, as a personality, has faded into the past. "His name is almost forgotten," Roger Burlingame remarks acidulously, "because our historians have been more interested in the heroes of politics and the battlefield than in those of industry."

Colt's Armory and machine shop. *(New-York Historical Society)*

He began as a bobbin boy in a textile factory, moved into a machine shop in Chicopee, Massachusetts, then went to the Collins Company, which made axes. There or earlier, as he heard and saw what was going on in the Springfield Armory and machine shops up and down the Connecticut River, he got the fever—make machines do the work of men—and production of axes in the Collins shop soared. He was forty-one when Colt lured him to Hartford in 1849. Six years later he had installed nearly four hundred machines to do the work men had once done.

Root veered from the road that Hall at Harpers Ferry wanted American technology to travel. He sought mechanization, which led to mass production, rather than precision, which led to interchangeable parts. Colt blustered when he bragged about the uniformity of his revolver parts. They "did not come close to being interchangeable," David Hounshell states. Visitors who came to the factory cared little about esoteric items like uniform parts. The machinery Root had installed fascinated them. Once in 1857 described what he saw: "Each portion of the firearm has its particular section. As we enter ... the first group of machines appears to be exclusively employed in chambering the cylinders; here another is boring barrels; another group is milling the lock frames; still another is drilling them; beyond are a score of machines boring and screwing the nipples. ... here are the rifling machines ... now we come to the jigging machines that mortice out the lock frames."

Samuel Colt died in 1862 at the age of forty-eight. Elisha Root, perhaps the first among the dirty-fingernail people to do so, became president of what was now a large, exceedingly profitable corporation. He died three years later, at the age of fifty-three.

21

The Sewing Machine

Roger Burlingame years ago noticed that Elias Howe's confrontations with the ''colorfully scandalous'' Isaac Merrit Singer had all the ingredients for a Hollywood blockbuster:

> One man [Howe] plods and suffers, nearly starves, loses his money, his wife, almost his mind and heart; another man [Singer] uses his genius to create crowds of eager prospects, makes an incidental fortune for himself but, lo! Enter the Law, justice is done, the unhappy victim on the threshold of starvation is rewarded and can sit back at last from his labors to profit from the infringer's work—then, in the end Fate, relenting again, admits the infringer too to a place in the sun and both die rich and happy.

Elias Howe was what the world calls a ''born loser.'' His father, a substantial gristmill owner, seems not to have discouraged the lad from leaving home early. The year 1835 finds Howe, then sixteen, working in a Lowell mill. Sometime later, in Boston, then a journeyman mechanician, married with three infants to feed, this ''Yuppie'' of the 1840s, eager to escape the blue-collar world, invented the sewing machine. This project evolved from one brilliant insight and two inventions. The insight, like that of McCormick, who saw that a machine could never duplicate the way a scythe reaped grain, revealed that no machine could replicate the way a woman sewed. The two inventions that made Howe's machine marketable were the following: First, putting the thread in the point of the needle, which, once it had punched through the cloth at right angles instead of obliquely as in hand sewing, would form a loop of thread on the underside. Second, as Burlingame puts it, "Seeing this loop on his experimental machine, Howe remembered the loom and knew that a shuttle carrying another thread could be made to pass through the loop so that when the

The historian James Parton's caption for this illustration reads: "The first contest between hand and machine sewing at Quincy Hall, Boston. Elias Howe with his machine on one side, and five picked seamstresses on the other. The victory for the machine was an easy one. At this time the journeymen tailors remarked, '320 stitches a minute at first trial. Death to sewing machines or death to tailors!'" (*New-York Historical Society*)

needle moved back again a lock stitch would be formed by two
threads.''

Howe does not seem to have thought of his machine for the
home or for housewives. He first showed it off at a clothing factory in Boston, where his challenge to five speedy seamstresses led
an umpire to say that ''the work done on the machine was the
neatest and strongest.'' Public apathy toward his invention led
Howe to London, where he expected to make his fortune ; a corset
maker bilked him out of profits, and Howe returned home penni-
less and a widower. He found at home that several people had
independently produced and were marketing their own sewing
machines, among them a song-and-dance man named Isaac Merrit
Singer. Singer had been reared in upstate New York, left home
when twelve, and roamed the Northeast for the next sixteen years.
He worked for a time in the theater, then drifted into shops where,
for one so flamboyant, he revealed an unsuspected empathy with
machinery. In 1850 he redesigned Howe's invention. His two crucial innovations were a metal ''foot'' that held the cloth in place
and a continuous feed that moved the material along with each
stitch. Despite the improvements, his machine clearly infringed
on Howe's patent, but Singer refused to concede this fact. Howe
took him to court, and early in 1854 the judge found in his favor.
''The public are indebted to Mr. Howe,'' said the court and then,
in a prescient sentence regarding precedence in American inventions, went on to say, ''A machine, in order to anticipate any
subsequent discovery, must be perfected—that is, made so as to
be of practical utility, and not merely experimental and ending
in experiment.''

Singer had to pay Howe $15,000 outright and a royalty on
every future machine Singer or anyone else manufactured. Singer's lawyer, Edward Clark, who relished industrial order, became
Singer's partner and pioneered the Great Sewing Machine Combination of 1856, which pooled the now numerous patents—they
amounted to more than eight thousand by the end of the century.
Anyone willing to pay a license fee could share in the pooled
patents. Howe received a royalty of $5 for every machine sold in
America and $1 for those sold abroad. After this settlement, more
than $2 million rolled in to the long-suffering Howe. Fortunately,
it came quickly, for he had only a decade to enjoy the new wealth.

The original Singer sewing machine. *(New York Public Library)*

He died in 1867 at the age of forty-eight. Singer lived on to 1875 and the age of sixty-four, but his name survives down to the present mainly because of his new partner, Clark, who channeled Singer's talent for public relations into the creation of a company that produced hundreds of thousands of sewing machines at a gigantic profit.

The sewing machine was probably the most complex item manufactured in America prior to the Civil War. Into it went ''pig-, bar-, and sheet-iron, iron and steel wire, bar- and sheet-steel, malleable iron, japan varnish, and power and machine supplies in general, woods for casing (largely walnut and poplar), besides a considerable range of other material.'' All of these components passed from the "tumbling room" to "annealing, japanning, drilling, turning, milling, grinding and polishing, ornamenting, varnishing, adjusting, and proving departments,'' ending, finally, in a mammoth assembling area. Subcontractors produced many of these parts, but the work of all was closely checked. A supervisor noticed on one order that "the throat plate sets in on the edge to such an extent that there was $\frac{1}{32}$ of an inch too much room.... When I discovered these I condemned them all.'' All parts called for close fits, and every machine when it left the factory was expected to be reliable and to run quietly enough to satisfy the most finicky housewife. A sewing machine was one of the most expensive single purchases in a woman's lifetime.

Once the patent pool had been created, the annual production of machines soared—from about 100,000 in 1860 to more than 700,000 a decade later—but it tended to center in three companies. The first, Wheeler & Wilson, produced machines by techniques pioneered in the two government armories. The three chief mechanicians—James Wilson (no relation to one of the owners), William Perry, and Joseph Dana Alvord—had been trained in armory practice, as it was called, in shops scattered about New England. They designed and built special-purpose machines that produced interchangeable parts. By 1862 they had created a plant capable of turning out 30,000 machines a year; production peaked in 1872 with 174,088 machines. The second company, Willcox & Gibbs, hired Brown & Sharpe, then a small job shop in Providence but guided by one of the few authentic mechanical geniuses of the day, Joseph R. Brown, to produce twelve special-purpose machines to make sewing-machine parts. Brown—he left the business side of the shop to Lucien Sharpe—had no experience in armory practice, but an employee who did helped to guide him. The order was slowly but eventually filled. When Henry Leland joined the firm in 1872 and soon came to head the sewing-machine department, the shop had become a factory with more than three hundred employees. Brown & Sharpe continued to supply Willcox & Gibbs with special-purpose machines until the 1950s.

I. M. Singer, the third company, took a different route. It ignored armory practice—neither Singer, nor Clark, nor any of their mechanicians knew much about it—and relied on subcontractors or on general-purpose machines to produce parts that had eventually to be filed by hand to fit. It concentrated on marketing. "A large part of our own success we attribute to our numerous advertisements and publications," said Edward Clark. "To insure success only two things are required: first, to have the best machines and second to let the public know it." Like McCormick, the company inaugurated installment buying and franchised retail shops and service centers throughout the country. It hired women to demonstrate the machines. Sales territories were tightly controlled from the central office.

Why, David Hounshell wonders, did the Singer company stick to old-fashioned production methods? For one thing, "Singer probably manufactured a sewing machine better in workmanship, that is, with closer fits, than those produced wholly with armory practice." Possibly, too, the company found armory techniques "unable to meet the standards of quality it desired." Finally, it was probably cheaper to hire men and women to do the finishing work than to invest in costly special-purpose machinery. Singer may have been technologically backward, but it was "able to sell sewing machines at five to ten times its cost of production" and to sell them "at the top of the price list for the industry throughout the nineteenth century."

22

Interchangeable Parts

The song-and-dance men—beginning with Whitney and continuing with Colt, McCormick, and Singer—filled the western world with claptrap about America's system of interchangeable parts. Colt spoke for them all when he described the system : "All the separate parts travel independently through the manufactory, arriving at last, in an almost complete condition, in the hands of the finishing workmen, by whom they are assembled, from promiscuous heaps, and formed into fire-arms, requiring only the polishing and fitting demanded for ornament. . . . they fit exactly each other, and any part to which they may be properly applied ; so in using these arms, new ones may be at once constructed out of broken ones left on the field of battle *without* altering the shape or size of the parts."

Colt and those who adopted his production methods turned out articles with uniform but not interchangeable parts. Why did he and the others ignore the achievement of John Hall at Harpers Ferry ? One reason was cost. Except in the clock industry, "where the required level of precision was generally very low," most manufacturers put aside the ideal of interchangeable parts because it was "prohibitively expensive." It would take years to pay for a machine that did what a skilled fitter with a hand file could do at his workbench for a meager weekly salary. The revolutionary weapon Hall created ultimately cost more than $2 million to develop, and "only the federal government could have financed such a massive undertaking," Merritt Roe Smith has noted. "That it did so over a forty-year period underscores the importance not only of capital but also longevity as a key ingredient in the evolution of complex technological systems. What the government provided, in addition to large infusions of money, was an ongoing bureaucratic organization with which the new technology—itself a bureaucratic phenomenon—could evolve." Costs aside, some manufacturers were not convinced that the interchangeable-parts system produced better products than those shaped and fitted for the market by skilled mechanicians. Singer, proud of its standards, ended relations with a supplier who turned out interchangeable parts because it "failed to produce machines of sufficient quality to meet the Singer officials' approval."

Technology had to make huge advances before the world of nuts and bolts was prepared to accept a system of interchangeable parts. Cheap and abundant steel, which holds its shape where iron does not, is needed; precision grinding machines; automatic machines that are "smart," as David Landes puts it: "They must know how far to travel, which is another way of saying that they must make the same pass time and time again. They must be neat, for cleaning entails deformation. And they must be tough, for any wear of the cutting tool will change the result. All these requirements, of course, vary in rigor with the margin of tolerance. The smaller the margin, the more precise and consistent the machines have to be." These tolerances got tighter and tighter as the century progressed. Then in the late 1860s came a pocket-sized, inexpensive micrometer that almost any mechanician could afford. "The micrometer made it possible for workmen to measure the parts they made with reasonable assurance that they were to size within 0.001 of an inch," Paul Uselding explains, adding that today "its inherent accuracy is usually between fifty-millionths to one ten-thousandth of an inch (0.000050 to 0.0001)."

America at last seemed ready for the proliferation of an interchangeable-parts system, but modern scholars find she was not. Uselding concludes that "it is doubtful that the diffusion of practical interchangeability . . . had proceeded very far in America by the close of the Civil War." Otto Mayr and Robert Post believe that even "in 1880 interchangeability remained unusual." Eugene Ferguson holds that "not until the twentieth century did interchangeable parts become a generally viable component of the American system." Even then many entrepreneurs still found it too expensive, and others were slipshod practitioners. Robert Howard found in a 1980 survey of private firearm manufacturers that "only Remington claimed some models of shotguns were completely interchangeable." Xerox almost foundered because suppliers could not meet its needs for interchangeable parts. The first marketed model of its machine had 1,260 components, all built by subcontractors. Xerox warned them that all parts must be interchangeable, but "they were simply not attuned to our exacting needs." When their parts were rejected, they "all but screamed their protests." An executive told them "we simply have to have better workmanship. . . . We must have copiers that are not only as near perfect as we can make them; *they must have interchangeable parts.* . . . This means every part will have to be manufactured to the tightest possible tolerances, tighter than anything we have so far attained." It took months to bring suppliers around to Xerox's standards.

It is easy to understand why, in an age of cheap skilled labor, when precise measuring tools were rare and special-purpose machines were expensive, entrepreneurs resisted the interchangeable-parts system. A more interesting question is why so many manufacturers lied in their publicity that their products were built with parts that could be interchanged. A generous explanation is that they did not know they lied, that they thought by producing uniform parts filed to fit they were actually turning out interchangeable parts. Another explanation, probably closer to the truth, is that they saw the words "all parts interchangeable" as a wonderful sales gimmick. To suggest to farmers, cowboys, and housewives that their machines came to them from the factory "untouched by human hands" offers a warranty hard to beat. It is a wonder none of the early song-and-dance men thought of that line.

23

When to Innovate, When Not To

Years ago during the Great Depression a story circulated about a country housewife who found a hobo at the kitchen door one morning begging for a meal.

"Are you willing to work for it?" she asked.

"Yes, ma'am," he answered.

"I have an easy job for you," she said, and showed him to a basement heaped high with newly dug potatoes. "Separate these into piles of large, medium, and small, and when you are finished come up for dinner."

An hour later the man reappeared in the kitchen and said the job was too much for him. "Why?" asked the astonished housewife. "It's those damn decisions, ma'am," he said and left.

Damn decisions, part of entrepreneurs' daily fare, might differ in time and from place to place but one—when to innovate, when not to—has been a constant in the history of American technology that all at some time have had to face. Often the fate of their business hangs on how they confront what can be a traumatic dilemma. Henry Ford clung to his Model T to a point that he nearly killed the company he had created. Andrew Carnegie's dictum—"It don't pay to pioneer"—continues to haunt entrepreneurs. "When I was at Oldsmobile," a president of General Motors once said, "there was something I learned that I've never forgotten. There was an old guy there who was an engineer, and he had been at GM a long time, and he gave me some advice. He told me, whatever you do, don't let GM do it first.''

Some entrepreneurs are ''inhibited by an aversion to the social consequences of technical changes,'' Habakkuk has said. ''Samuel Crompton, for example, intended the mule to simplify the labour of the spinner in his home and increase the volume and the quantity of his output. The application of steam to gigantic mules of 600–700 spindles was no part of his original plan, and 'no man lamented more the changes thus brought about in the life of the people.' '' Too often, David Noble writes, historians portray ''technological development as an autonomous and neutral technical process, on the one hand, and a coldly rational and self-regulating economic process, on the other, neither of which accounts for people, power, institutions, competing values, or different dreams.'' Such an approach ''begs and explains away all important historical questions: The best technology? Best for whom? Best for what? Best according to what criteria, what visions, according to whose criteria, whose visions?''

From scores of examples that might be offered let the reactions of two men to innovations suffice to illustrate the complexities involved. Thomas Gilpin faced the great decision in the early 1800s and John D. Rockefeller toward the end of the century. Both men normally welcomed technological advances that promised to increase profits, and then suddenly, almost overnight, each had to deal with an innovation that could wreck the enterprises they had spent years creating.

Gilpin and his brother Joshua early in the nineteenth century had erected the first mill in America that manufactured machine-made paper. It turned out in a few hours what it took a week or two to produce by hand. For several years the Gilpins had a monopoly on machine-made paper that looked unbreakable. They had stolen the design for their machine from John Dickinson in England, and then to assure that it stayed in running order they seduced Dickinson's foreman to supervise their mill on Brandy-wine Creek. Then one day in 1817 or 1818 Coleman Sellers and his ten-year-old son George arrived at the door with startling news. A mechanician in New Jersey had invented something called a squirrel-cage cylinder that simplified and greatly reduced production costs of machine-made paper. Sellers had improved the cylinder and said he could market it for about $160. He estimated that in a short time "the cost of putting up and starting a simple squirrel cage cylinder machine with wooden press-rolls [would be] from $500 to $600," a fraction of what the Gilpins had spent to build their own machine. Years later George Sellers reported Thomas Gilpin's reaction to this news: "I recollect that when my father sketched and explained this simple cylinder to Mr. Gilpin that he seemed perfectly incredulous as to the possibility of consolidating a film of pulp sufficiently to couch without the V-shaped trough and exhaust pumps, but at the same time he showed considerable uneasiness; for, said he, could such a simple cylinder and machine be made to work at all it would be a dangerous competition to his expensive machines." Gilpin's uneasiness soon proved justified. Within a decade the squirrel-cage cylinder, along with further innovations purloined from England, had revolutionized the paper-making industry in America and, said George Sellers, "the rapid change from hand-made to machine-made paper had fairly set in." Thomas Gilpin swung with the tide. He modernized his mill, and it stayed profitable, but his monopoly on machine-made paper was forever broken.

Sometimes even when acceptance of an innovation appears to be inevitable an entrepreneur normally attracted to carrying out a routine process in a new, cheaper, more efficient way will still resist it. John D. Rockefeller always thought of himself and his company as "progressive," to use one of his favorite words. Originally, for instance, like all refineries, Standard Oil hauled the crude in barrels from the fields to its refineries in Cleveland, Pittsburgh, Philadelphia, and elsewhere. Then along came the tank car, which did the job for less than half the cost; Standard quickly adopted the innovation. The company had been one of the first to install pipelines from well sites to rail heads over the violent objections of the draymen who had once done the work. But when in the 1870s independent competitors in Pennsylvania's Bradford oil fields conceived the idea of moving crude by long-distance pipelines directly to the refineries, Rockefeller was repelled by the innovation, though he knew that such lines would cut transportation costs to a fraction of what the railroads charged even the favored Standard. He argued that long-standing secret alliances with the roads had helped to build the Standard Oil Company into a bountiful empire; those agreements must be honored. "We do not want to invest any money in transportation enterprises in competition with the roads," he said. Past favors and old agreements had trapped him into rejecting out of hand the greatest technological innovation that had come along since he entered the oil business.

He had a further reason for resisting long-distance pipelines—they could possibly annihilate his delicately balanced, loosely knit empire. Eastern refineries might continue, for their markets were close by in the populous coastal cities; but inland centers like Pittsburgh and Cleveland, forced to ship their refined oil by railroad (it was mistakenly assumed refined oil could not be pumped long distances through pipelines) would wither away. Rockefeller saw no choice but to fight the pipeline revolution to death.

Standard battled the pipeline every foot of the way from its source in northwestern Pennsylvania down to Williamsport, where the Reading Railroad, no ally of Rockefeller's company, was to carry the crude to independent refineries in Philadelphia and metropolitan New York. It bought up rights to a broad north-south strip of land, hoping to block the line as it pushed southward. The pipeline company found holes in the wall large enough to poke its line through. Standard planted stories in the press suggesting that a farmer who leased his land to the pipeline risked having his fields ruined by oil leaks. It tried to buy up all the "black gold" produced by independent drillers in Pennsylvania who seemed inclined to ship their crude by pipeline. It flooded tank-car builders with orders in order to keep the Reading from accumulating its own fleet.

Colleagues told Rockefeller that the pipeline would succeed, that Standard should cease to resist it, and that inevitably it would have to accept the revolutionary innovation. For one of the few times in his life John D. willfully deluded himself and blinked away the inevitable. He said that the two pumps designed to shove oil over mountains some 2,500 feet high would never do the job. Shortly before the pipeline's cocks were opened he was still "not a little skeptical" that it would work. Three weeks later oil began to flow from the Bradford fields into Williamsport. It poured in at the rate of 250 barrels an hour, 6,000 barrels a day.

Instantly Rockefeller reverted to form. Standard went into the pipeline business and within a few years had a near monopoly on the new way of moving crude from the fields into the refineries. But Rockefeller never boasted of the eventual victory; the campaign had not been part of that progressive, upbuilding process he liked so much to talk about. Forever after he tried to obscure details of his embarrassing battle against a major technological innovation. As an old man he sat one day shaking his head over a passage by a friendly historian about the ''great obstacles'' Standard had raised against the pipeline pioneers. ''I don't see any reasons to make it apparent that the Standard Oil Company did all they could to prevent it,'' he said, and that was that.

VI

A Diffusion of Knowledge

24

By Print
and Word of Mouth

In the novel *Red Storm Rising* the author, Tom Clancy, describes how an American intelligence officer divines from scattered hints that Russia is planning to launch a war against Europe. The first hint comes from a little-noticed report that the Russian army has courtmartialed and shot four colonels for relatively minor infractions. Why? Then he learns from a Navy friend of a sudden and inexplicable shortage of car and truck batteries in Russia. A satellite shot shows submarines normally at sea now clustered in home ports. Another seemingly innocent photo indicates that the size of some private plots on a collective farm have been expanded. Slowly these meaningless pieces of information form a pattern that convinces the protagonist Russia is about to go to war: the colonels' deaths told the army that discipline was about to be tightened up; batteries were being diverted from civilians to the submarine fleet; reserves were being called up and those left behind had to produce more food.

Early nineteenth-century mechanicians and entrepreneurs often picked up hints about British technology from immigrants and smuggled machines, but equally often those hints were as meaningless as those Clancy's intelligence officer first stumbled across. Americans resorted to espionage, with indifferent success. They interrogated informed immigrants as diligently as defectors are queried today. But it took time for a comprehensible pattern to form. Several of Britain's key textile machines that found their way to America ended up useless because no one knew how to assemble or to operate them. Sometimes a mechanician knew about a particular machine but not its role in a sequential line of machines. (Moses Brown found himself in that position until he lucked onto Samuel Slater.) The ways knowledge spread then differed little from today, with a single large exception. Defections were encouraged by the American government and by state governments; the bait of £100 published in a Philadelphia newspaper drew Slater to America. The large exception was printed matter.

Clancy's extraordinarily informed novel about NATO, Russian, and American military equipment has been drawn largely from what is in the public domain, printed matter. Early ninteenth-century America lacked a similar source. Anthony Wallace, speaking of the textile industry's troubles, says that "at the outset [the entrepreneurs needed] printed literature, but there was very little available in 1821, and that little was inadequate." Not until the 1850s could entrepreneurs draw on published literature to inform themselves of the latest technological innovations.

Word of mouth, then, was in the early years the main way technological innovations spread about the country. Merritt Roe Smith has resurrected the little-known facts about Nathaniel French, whose career as a mechanician "exemplifies the roving tendencies of early American mechanics." He arrived at Harpers Ferry in 1821 with considerable experience, for John Hall hired him as a pattern maker and machinist at the high wage of $1.75 a day. He left six years later, apparently drawn away by a better salary, to work for Simeon North, the private firearms contractor in Middletown, Connecticut. Back to Harpers Ferry, then, in 1831, he moved on to the Springfield Armory. "The implication is clear," says Smith: within a few years "Nathaniel French had plied his trade at three leading arms-making establishments. The mechanical information he carried from one armory to another cannot be dismissed lightly, for here was fresh know-how and the ability to execute such work embodied in one and the same person."

French's perambulations call for further comment. Note that he moved within a constricted circle, from one firearms establishment to another, all working to produce an effective interchange-able-parts system. He, like most mechanicians, for all their roaming, went from the familiar to the familiar. True, as they wandered they picked up new ideas that expanded previous experience, but only to a limited degree. George Escol Sellers, a knowledgeable, bright young mechanician, had traveled widely among shops in the Middle States, but when he went to England in 1832 he knew nothing about, had never heard of or been exposed to, the interchangeable-parts system being developed within federal armories. Until the printed word caught up in the late 1820s and early 1830s, American mechanicians were parochial, their knowledge and experience mainly limited to local areas. Regional variations pervaded American technology. Philadelphia built fire engines one way, New York another. New Englanders adapted a slide lathe stolen from Britain to suit themselves, Brandywine entrepreneurs chose innovations that suited their needs and the markets they had at hand.

The essays Oliver Evans published describing his automated gristmill and his high-pressure engine inaugurated a new era : they presented to the nation, not particular parts of it, two American innovations in clear, forceful language a layman could understand. The Franklin Institute, organized by Philadelphians in 1823, tried to duplicate Evans's achievement in a different way. Its journal, edited by Dr. Thomas P. Jones, a sometime lecturer on science who in his spare time hung around the Sellers workshop, used language "intelligible to common workmen" to tie science with what went on in the workshop. It published lists of the latest patent applications with commentary by the editor, and such articles as how to design an efficient waterwheel and how to counteract the "increasing scarcity and price of wood" with the use of anthracite coal, which gave off a hot flame but was hard to ignite. The institute staged annual exhibitions to encourage and publicize innovations and to discourage "the predilection Americans had for imported goods," notably those from Britain. By 1838 these fairs were drawing forty thousand visitors. Prizes were awarded to entries that solved practical problems—a gadget to keep sparks from escaping out of a locomotive smokestack, the best coal-burning domestic stove "the price not to exceed ten dollars," the best lamp for lighting city streets. By 1840 "the

Institute's example had furnished the pattern for industrial fairs in most major cities throughout the nation,'' Bruce Sinclair, its historian, has said. ''Indeed, attending such events became a great American pastime.'' By the time of the Crystal Palace Exhibition in London in 1851 ''Americans had a quarter century of experience with industrial exhibitions.'' These fairs—and since 1825 there had been on and off national versions of them in Washington—''were an important vehicle for the inculcation of faith in industrial progress and technological advancement and for emphasizing resource development, mechanization, and inventiveness. The ingenuity which surprised visitors to London in 1851 was something Americans were accustomed to seeing.''

These fairs, while not unique to America, exemplified a characteristic that *was* unique—its ''open shop'' policy, as it has been called, meaning the lack of secretiveness, the free-and-easy exchange of ideas among mechanicians and entrepreneurs. George Escol Sellers became aware just how American this openness was when he visited England in 1832.

25

An American
in England: 1832

George Escol Sellers was twenty-four when he embarked for
England in 1832. His father hoped a sea voyage would speed the
young man's recovery from an attack of cholera, and he also
wanted him "to see and learn all that was accessible in the way
of civil and mechanical engineering." Friends in England wrote
that "access to many mechanical works was difficult, and, in some
cases, impossible for Americans," particularly the workshop of
Bryan Donkin, which was "absolutely closed to all foreigners."
Donkin had pioneered a revolution with his Fourdrinier continuous-web paper-making machine. Most paper in America was
made by hand, and it took more than a week to fill a customer's
order. With one of Donkin's machines "an order may be taken
for any desired thickness, size and finish and within an hour from
starting the paper machine ... packers may begin putting up the
order." Coleman Sellers designed and built machinery for one of
the few automatic papermills in America, owned by Joshua Gilpin. Both men were almost obsessively curious about Donkin's
machine, and they gave George Escol the seemingly impossible
task of finding out all he could about it.

Young Sellers landed at Liverpool, and to get acclimated to the
English world of mechanicians he headed for London by a roundabout route through workshops and mills in Manchester, Birmingham, and Sheffield. Doors opened along the way because he was

accompanied by an influential family friend, but also because he was an affable young man, a good listener, no chauvinist, and knowledgeable enough to talk shop with the mechanicians he met. He reacted with mixed feelings to much that he saw in the Midlands and later in London. Sometimes the workships were gloomy dens. One mill he visited was surrounded by a moatlike race over which loomed a high wall punctuated with small windows. It "gave the appearance of a fortified place or jail. It was hard to realize it was a paper mill, so unlike our light and airy mills." In one shop a guide boasted of the huge cylinder then on the boring machine. "I must confess to a feeling of great disappointment," Sellers said after viewing it, "for the cylinder struck me as a mere pigmy compared with the cylinders of the North River and Long Island Sound boats." "The amount of labor and useless cost of work done on lathes not calculated for the work they were doing" astonished him. He could hardly believe that the slide-rest lathe the great English mechanician Henry Maudslay had invented years ago remained virtually unchanged. Back in Philadelphia Moses Tyler and Isaiah Lukens had built a string of improvements into the lathe, "which for solidity and firmness so increased the amount of work that the Maudslay rest was laid aside and the lathe transferred to the pattern shop as a light-running wood lathe long before I went to England." Sellers wished English mechanicians could see "the festoons of great wrought-iron shavings" that came from American lathes. "It would spur them on." Yet he could only admire "the workmanship . . . of the highest possible character" that came from obsolete machinery. The beautiful work produced with awkward hand tools astonished him. "Large flat surfaces . . . came from the cold chisel almost as perfect as they now do from our planers. This was explained on the principle of division of labor, men working a life time with hammer and cold chisel. When I took in my hand their heavy, short, stubby cold chisels, their short, clumsy, broad-faced, short handled hand hammers, I felt it would be impossible for me to handle such tools, with any prospect of approximating their results."

While in the Midlands Sellers toured three workshops that used Donkin's machines. "The beauty of these machines and great perfection in performing their work made me more desirous than

ever to become favorably acquainted with Mr. Donkin.'' From
the first two owners he collected vaguely phrased letters of introduction to Donkin, who, he gathered, was a crusty, secretive old
man. At the third shop the owner said those letters would never
do, and he sat down to write one of his own.

> He simply told Mr. Donkin ... that I was engaged in America
> in the same line of business that he was; that he believed a free
> interchange of ideas would be mutually beneficial; that he could
> assure him that I would not intrude or be inquisitive into any
> matters that he was not disposed to communicate. He then ad-
> vised my going to Mr. Donkin without anyone with me, and in
> handing him the letters to be careful so to place them, that he
> would be likely to look at the others before opening his; he told
> me of many of the little peculiarities of Mr. Donkin and told me
> to observe him closely as he read the letters.

Sellers knew the biography of Donkin before he met him. Some
thirty years earlier two British stationers, Henry and Sealy Fourdrinier, had bought half-interest in a continuous-web paper-making machine designed by a Frenchman. It was ''in principle
correct but in detail so imperfect that the machine ... was an
absolute failure; but Donkin's brains and skill made it a perfect
success. Not at once, but by long years of perseverance, improvements and additions [he] produced the splendid perfect machine
that is known now as the Fourdrinier Long Web Machine, a mis-
nomer; it should have been a Donkin. Fourdrinier's part was like
that of Boulton to Watt, capital and confidence.''

The opening interview took a predictable course. The third letter of introduction impressed Donkin. ''I am glad you have not
come under false colors,'' the old man said after reading it, ''as I
am sorry to say mechanics have done.'' Perhaps to test his young
visitor he aired his bitterness about America. ''My ideas have
gone to America in a machine I sent there to fill an order, and I
learn they have already been copied.'' Sellers did not rise to the
bait. Instead he talked about ''our mode of putting up the cylinders, the machinery and tools we used.'' Donkin listened but said
nothing. Sellers said he ''had been surprised at not finding the
direct-acting guillotine paper cutter in use in England.'' It was
used everywhere in America by bookbinders. Donkin perked up,

asked to hear more. Sellers explained how it worked—"the pile of paper to be trimmed being pressed tightly together, the knife goes through it like shaving wood." The inventor held an American patent and Sellers's father paid "the patentee a royalty on each machine."

Donkin must have sensed he had at last met an honorable American. He took young Sellers into the erecting shop "where stood, certainly, the finest specimen of workmanship that I had seen in England." There the tour ended. Donkin explained: "the tools and various machines and appliances I employ in their construction have been the work of almost a lifetime, and I hope you won't take amiss my unwillingness to exhibit them."

Sellers returned to his lodgings to find waiting there Joseph Saxton, his old friend from Philadelphia, now learning what he could about English ways of doing things.

"Did he show you through his shops?"

"Only his erecting shop."

"I told you so; they are closed against all foreigners, particularly Americans."

"But the end has not come," said Sellers.

Nor had it. Several visits with Donkin followed. During them trust, respect, affection even, developed between the old man and the young one. Eventually Donkin did what he had never done— gave a personal tour of his workshop to a foreigner. A long afternoon of talk had led up to the tour. More than half a century later Sellers recalled how he had tried to explain America to one of England's greatest mechanicians. He had showed Donkin a squirrel-cage cylinder designed by his father as an improvement over a similar cylinder then used in paper-making machines. It was that, in the excerpts that follow from Sellers's reminiscences which set Donkin off.

He asked what we got for a complete squirrel cage cylinder, naming a size. On my giving our regular price, he promptly said that he could not compete, pay freight and duty, for our price was less than he got at his works.

This I felt to be my opportunity, so I explained our mode of putting up the cylinders, the machinery tools we used. He listened very attentively, but made no comments. I could not but admire his extreme caution and reticence as to his modes. At the same time his evident eagerness to learn what others were doing amused me, and I felt much like a man in the hands of an interviewer of the present day.

Great and successful inventor as he was, and one who had done so much in perfecting whatever passed through his hands, and who was certainly the most progressive machinist I had met in England, yet he seemed to labor under false impressions, and not clearly to understand the condition of things that led to such rapid advances in mechanical pursuits in the United States. He made notes of the wages we paid for skilled labor and such sort of crude materials as I could give him. Then he came back to the squirrel cage cylinder, and said he could not see how we could afford them at the price I had named. As I had myself made many of them, I went fully into detail, and seemed to satisfy him that the higher wages naturally led to mechanical contrivances, and that, in the case of the cylinders, they were of the simplest possible kind, and yet as labor savers that portion of the cost was reduced below the cheaper labor in England; that, in the crude materials, the iron and bars, the saving was made in proportioning the parts to the work they had to perform, the American cylinder not weighing over two-thirds that of the English.

Laying on his table I noticed what appeared to be samples of pliers, nippers, and a few such like tools. I picked up a pair of pliers, and remarked that it looked like an American tool—not so clumsy as those I had seen in use in shops I had visited.

"Strange," said he; "they are samples from Stubs, of Sheffield, and they are sent as the American pattern," and he supposed, were being introduced under that as a distinctive name.

He seemed greatly surprised when I told him they were fairly entitled to be called the American pattern; that the brothers B. & E. Clark, of Philadelphia, watchmakers, in addition to their watch and clock repairing business, kept a supply store of watch and clock makers' materials, including tools; and in my earliest recollection they were the only parties in Philadelphia that kept on sale Stubs steel files, etc. They were fine workmen and ingenious men, who either altered English tools or made those they used of such form and proportions as they found best adapted for the general work they had to do. Samples of these were sent to Sheffield to be duplicated, and for a considerable time they were the only parties who kept them for sale; but they had spread until they became universally adopted. That in Birmingham I had noticed the same thing taking place in general hardware—the class being made for the American market materially differing from that for home consumption, being generally lighter and more elegant in form. That I had learned that in every case the change had been made to conform to patterns or drawings sent from the United States.

He spoke of the feverish state of excitement among his best skilled labor, owing to the glowing accounts they received from brother workmen who had emigrated, and he asked me as to their real condition. Men, he said who, on the English plan of division of labor, were only perfect on a single branch, he did not believe it possible could find constant employment on that —in a comparatively new country.

I told him that he must bear in mind that America's start in mechanical art was at the point England had reached and without her prejudices. That the men who at home would resist the introduction of labor-saving machinery were glad to accept such as they found in America, as by it they were enabled to turn their hands to general work as it offered. I reminded him of the English prejudices that years before had led to the riots that destroyed the nail-cutting machines that Samuel R. Wood of Philadelphia was endeavoring to introduce in England. Wood was a member of the Society of Friends and non-combative, and he left England in disgust.

I said it would be impossible to estimate or realize what the rejection of the cut nail had cost England. Its invention in America filled a vacuum and was almost a necessity, not only as to first cost of the nails but as great labor-savers in carpenters' work; that I had noticed that in England every carpenter had in hand either brace and bit, gimlet or brad-awl, according to the work he was doing, for without them the square, uniformly tapered hand-made wrought nail was the best possible form that could be devised to split the wood it was driven into, without first boring a hole to receive it; that its tapered form, if not driven through and clinched, would lose its hold on the least starting back—still they continued in common use; that on watching the joiners at work, I believed I was safe in estimating that for every English nail driven, the use of the American cut nail would drive four or five. That in patternmaking shops I had seen the wrought clout in use by having its head flattened edgeways by a stroke of a hammer, and then it made a ragged hole to be filled with wax or putty.

Mr. Donkin smiled as he said, "I have long been using in my pattern shop the American cut brands"; then he must understand the point; but I would give another instance of the fixed ways and prejudice of the old country that kept back improvements.

Mr. E. R. Sheer, a pianoforte maker of Philadelphia, in fitting work where wood screws had to be withdrawn and again driven in the same holes had found it difficult to make the common square-end English wood screw enter and follow the thread cut by the first insertion; he had mounted a clamp chuck on a foot lathe that would grasp the shank of the screw, then with file and chasing tool he tapered the end of the screw like that of a gimlet. He had given me several of these as samples, with the request that when in Birmingham I would induce some good screw maker to fill a considerable order of gimlet pointed screws. I had gone to the makers with a prominent shipper of hardware through whom they received most of their American orders, and we had failed to induce any one of them to fill the order; they and their predecessors had always made wood screws as they were then doing, and they would have nothing to do with such new-fangled notions.

After this monologue ended Donkin took his visitor through the workshop. Sellers was awed by what he saw. A giant, engine-powered lathe "certainly, for solidity and fine workmanship, came nearer to the lathes of the present day than anything I had previously seen in England." A further surprise awaited, and here again Sellers must tell what transpired.

As we were going through the shops, a clerk handed Mr. Donkin a letter that had been brought in haste by a special messenger: he glanced over it, asked me to excuse him for a short time, calling on the room foreman to show me around during his absence....

Mr. Donkin soon returned with the open letter in his hand, and said to me, "Here is a case in point, showing the value and importance of, as far as practicable, making all parts of machinery interchangeable. Mr. _____ has met with a serious accident to his Fourdrinier machine. The carelessness of an attendant allowing a tool to slip from his hand caused a break, that before the machine could be stopped was carried forward, doing serious damage to other parts of the machine; a messenger with a conveyance and the request that I lose no time in sending workmen with tools to make repairs. He has fortunately given in his letter a full detailed description of damage done, hoping that by so doing I would, in a measure, be prepared and that he would not be obliged to have his mill shut down for more than three or four days." He then added that in the short time he had left me he had dispatched a competent workman with duplicates of all the broken parts, and that by midnight he had no doubt the machine would be in running order. He spoke of having for years made a study of the practicality of making all parts of his machines of uniform size and shape, and having the work systematically done to rule by templets and fixed gauges. The key seats in light shafting were milled, but for heavy shafting and gearing the cold chisel and file were still doing the work.

At the noon hour, when the machinery stopped, I was taken into the storeroom, in which were arranged all the various parts of the Fourdrinier machine, with the exception of the frames, press rolls, and drying cylinders. It was from this room that the ready-made duplicates to replace the broken parts had been sent.

I would here note that 54 years ago this was the first instance I had seen where making the component parts of machinery interchangeable had been reduced to an absolute system, that is now so universally practiced by first-class machinists.

Sellers ended his account with words that could serve as an epitaph for Donkin: "He was the most advanced mechanical engineer of his time, and it is to his inventive ability, zeal and persistent application... that the world is indebted for... producing the self-acting endless web paper machine in such perfection by the year 1832, that in the 54 subsequent years no essential changes have been made, and now the great bulk of the paper of the world is produced on machines substantially as they came from his brain and hands at that early period.''

26

Brown & Sharpe

Of several early nineteenth-century New England tool shops—
Pratt & Whitney, Robbins & Lawrence, the Ames Company,
Brown & Sharpe—the Ames Company seems at first glance to fit
best into any account of the diffusion of knowledge among mechanicians and entrepreneurs. It was founded in 1834 by Nathan
P. and James T. Ames and pioneered the production of "a standard line of machine tools to the general public." Coleman Sellers
& Sons in Philadelphia was an early customer. Ames equipped the
Robbins & Lawrence factory in Windsor, Vermont, the first private firm to manufacture rifles with interchangeable parts. They
also sold abroad to Britain, Spain, and Russia. "Considering their
varied engagements for textile machinery, mining equipment,
millwork, cutlery, swords, small arms, cannon, bronze statuary,
and other metal castings, the activities of the Ames brothers aptly
illustrate how early tool builders served as key transmitters in the
diffusion of new skills, and techniques to technically related industries," Merritt Roe Smith writes; but, he adds, the firm was
"not known for innovations." By contrast Brown & Sharpe was,
mainly because of its presiding genius, Joseph R. Brown, and the
people he attracted to work with him.

Once again two individuals of later fame, one a mechanician
(Brown), the other an entrepreneur (Sharpe), come down to us
through a mist. Joseph R. Brown, the son of a clockmaker, apprenticed under his father; as a young man, he opened his own
shop in Providence, Rhode Island, and brought in Lucien Sharpe
as an apprentice. The two became partners. Sharpe was unique:

he had a talent for business, but his training in the shop made him "one who thoroughly understood, appreciated, and encouraged the work Brown was doing." The business began modestly. The shop made and repaired clocks and watches and precision instruments for local firms. In some way it acquired a reputation beyond Providence, for late in 1857 James Willcox, a businessman in Philadelphia, had heard of them and gave an order for twelve sewing machines designed by his partner, James E. A. Gibbs. Brown & Sharpe had the equipment at hand—two engine lathes, two hand lathes, an upright drill, two planers—to produce the machines if castings were subcontracted and all parts hand-fitted. They chose, instead, to make a large leap into the unknown and adopt "armory practice," about which they knew nothing, to make the twelve machines. The gamble involved huge risks—the Willcox & Gibbs machine had not been proved on the market—and the partners had no War Department to underwrite costs. Sharpe soon admitted that setting up a production line "has taken much longer than anticipated" and "the tools proved to be three or four times as expensive"—eventually ten times as expensive—"as was contemplated by us at the commencement, though they will doubtless be cheap in the end if many machines are manufactured." The goal, as it was to be through the company's history, was to produce "perfect work." Fortunately, Willcox kept the faith; the order for twelve blossomed into one for a hundred machines. By the end of October 1858, eight months after beginning work, the firm had fifty of the hundred being "finished and put together." Brown & Sharpe had gambled on sewing machines "to give us business enough for these tools for two or three years at least; else it would not be good policy to increase our facilities as much as we have." (The firm continued to turn out Willcox & Gibbs machines until the 1950s.) More important, the ingenious machines Brown had devised led Brown & Sharpe into the machine tool industry. "The company reaped more than financial profits from sewing manufacture," David Hounshell notes. "In addition to leading the company into machine tool manufacture, the Willcox & Gibbs business supplied Brown & Sharpe with an opportunity to test the tools it marketed and thus kept the company conscious of the real needs of manufacture in such vital elements as precision, speed, and ease of operation."

As the eminent historian of American machine tools, Robert

Woodbury, has remarked, Brown's machines "were not restricted
and specialized tools for only making sewing machines, but were
conceived in such broad terms that they became of first-rate importance in all machine-shop work." An early creation was the
"automatic linear dividing engine," which produced accurate
gears. Later came the "turret screw machine," then the "universal milling machine" (1861) and the "formed cutter"(1864).
These last two, like most machines Brown devised, "were developed in a form sufficiently general to be of wide use for many
other purposes." Brown's machines not only sold well but were
widely copied throughout the western world. His triumph, the
"universal grinding machine," was shown at the Paris Exposition of 1867, when Brown was fifty-seven. "It marked at once the
culmination of the development of the precision grinding machine
up to that time," Woodbury says, "and it also introduced the
basic form which the cylindrical grinding machine was to have in
the future." Henry Leland, a foreman in the shop when Brown
was creating the grinding machine, always considered it "Mr.
Brown's greatest achievement.... In developing and designing
this Machine he stepped out on entirely new ground and developed a machine which has enabled us to harden our work first and
then grind it with the utmost accuracy.... This in my judgment
is one of the most remarkable inventions and too much cannot be
said in its praise, or in acknowledgment of Mr. Brown's perseverance, wonderful initiative and genius."

The presence of Leland in the shop—actually a factory when
he arrived in 1872—attests to the self-confidence of Messrs.
Brown and Sharpe: they did not fear talent. Of a score of outstanding mechanicians who worked for them, three are singled
out from the panoply. The first is Frederick W. Howe. He had
roved widely among machine shops before Lucien Sharpe opened
a correspondence with him in 1859. He had designed a miller that
"was a very creditable piece of design embodying a number of
important advances," according to Woodbury, who prefers to
credit Brown for the ultimate universal miller that the shop marketed. Hounshell disagrees. Howe, he says, though employed elsewhere at the time, collaborated with Brown "to design a machine
to mill the grooves in twist drills, which resulted in the so-called
universal milling machine." Howe joined Brown & Sharpe in

1868. "His major contributions to the company had already been made," Hounshell says, "but he superintended the design and construction of Brown & Sharpe's new factory in 1872." Leland said later: "I felt then and I believe now that their new plant was, and for a long period in time remained, the finest of its kind in the world."

Henry Martyn Leland, reared on a New England farm, came to Brown & Sharpe at the age of twenty-nine with wide experience. During the Civil War he had worked at the Springfield Armory and after the war had shuttled among a variety of shops. Brown & Sharpe's management exposed him to all sides of their operation; during one stretch he worked closely with Brown, then creating his universal grinding machine. Leland, in time settled into the sewing-machine department, which he eventually, over much opposition, totally reorganized. His talent, as Hounshell puts it so well, lay not in creating machine tools but in the process of manufacture "rather than the building of any particular product or tool." He saw manufacturing as "an art," but of what did that art consist? He tried in old age to explain:

> The art of manufacturing consists first in knowing thoroughly what one is to make, its materials, its parts and their function; and second is the ability to select the most advantageous machines and tools for speedy operation, fewest cuts and fastest feed. Yet nothing is more important as far as shop expense is concerned, and as far as having the completed machine perform its functions properly, than judiciously placing the tolerances or limits. One *must* study and learn the work that each piece is to perform and adapt the operating tolerances to match. There is reason for fineness in some work, while to grind off a thousandth of an inch on others is ruinous to profits; and it calls for the soundest judgment to determine and set the limits. Then if one piece can be made true to the requirements, a million can be made in the same way and they will be exactly alike.

Leland left Brown & Sharpe in 1890. An incident that helped speed him on the way arose over a hair clipper he wanted the company to make. Sharpe—Brown was dead by this time—resisted the innovation, saying that no market for it existed. Leland found the market, and the factory began producing three hundred clippers a day. ''For this I received a 'thank you' and fifty cents a day more in my pay envelope,'' Leland recalled in old age. ''That was one of the times I thought I ought to quit making other men rich and go to work for myself.'' He soon quit and ended in Detroit heading his own machine shop, Leland, Faulconer & Norton Co. The ''Norton'' was Charles H. Norton, who had been enticed away from Brown & Sharpe. He apparently did not get along with Leland and Faulconer; his name vanished from the logo in 1893, and he returned to Brown & Sharpe.

Now that he was in the business of making parts for automobiles, Leland saw a flaw in Brown's grinding machine. It was, he said, too light. ''This fault was almost characteristic of his designing, and it is not surprising when one remembers that he was originally a watch maker. Then he made the small tools, vernier calipers, etc. . . . then sewing machines and a large line of milling machines and other work equally light. Since he had only experience on light work, it would be reasonable for him to design his new machines very light.''

Norton, too, must have been convinced by the Detroit experience that Brown's grinding machine, as he put it, ''was so light and its construction so frail, that it would be impossible to do work, commercial work, with any success, except by great skill and great patience.'' Back at Brown & Sharpe he designed a machine that could grind large pieces of metal precisely. His supervisor told him ''there was no need for such a machine.'' Norton soon departed and produced a machine that could shape in fifteen minutes automobile crankshafts that ''had previously required five hours of turning, filing, and polishing.'' Woodbury summarizes his achievement: he had shown ''how hardened steel parts of substantial size could be produced rapidly and cheaply, and with the necessary precision and finish—by grinding.'' His departure forced the Brown & Sharpe management to realize that a new age had arrived with the autombile in the world of nuts and bolts. To survive it must adjust. The fact that the company exists today shows that it did.

VII

Science Steps In

27

The Franklin Institute

Beginning in the seventeenth century, science and technology went their separate ways, and seldom did they communicate, though each could have learned much from the other. When an astronomer suggested that a clock pendulum oscillated faster in winter than in summer, the Dutch physicist Christian Huygens rejected the idea that temperature had anything to do with the change, even though clockmakers "unhampered by theory, knew better." Leonhard Euler and other mathematicians worked out "a mathematically derived theory of the most efficient profiles of meshing gears" decades after mechanicians had solved the problem empirically. The Franklin Institute was created in Philadelphia in 1824 to bridge this gap between the tribes. By then the mechanicians saw, as Oliver Evans had seen years earlier, that "they lacked a publication in which issues of importance to the profession could be debated and new information of general concern be communicated," Anthony Wallace states. They also realized that they

lacked a formal method of proof that went beyond the claims of experience and authority to be comparable to the experimental method of science or the logical methods of mathematics or theology. They had no way of conclusively demonstrating the validity of a technical proposition, such as "Overshot wheels are more efficient than breast wheels." And, finally, they did not have a system of formal education, which could standardize mechanical training and relate it to useful information in the sciences.

The founders of the institute hoped to fill in all of these lacunae.

The Franklin Institute was founded at a time when mechanics' institutes, then proliferating in Great Britain, were spreading to America. For the most part they were created and run by workingmen; they cut the mechanician apart from the mainstream and left him once again isolated in his separate world. Academies of Arts and Sciences scattered about the country, like the American Philosophical Society, were elite bodies, composed of gentlemen who liked to dabble in science; with few exceptions, social standing as much as an interest in science determined membership. The conception behind the Franklin Institute was unique: it was to weld gentlemen of Philadelphia and mechanicians into a body in which for the first time the dirty-fingernail people would be treated as equals. Hitherto this mixed breed of men "were to be found laboring in their useful avocations, without the stimulus of a public acknowledgement of their importance as a class of citizens, without any concert of action or interchange of experience." The Franklin Institute was created to alter this deplorable situation.

The institute in the beginning offered the usual vague promises—"to advance the general interests" of mechanicians and entrepreneurs "by extending a knowledge of mechanical science" —but it was also specific. Its goals were to examine and comment on all new inventions offered to the patent office; hold public lectures on science and offer classes "to provide instruction to workingmen in the principles of science"; publish a readable journal literate mechanicians could comprehend; and establish a mechanical drawing school, something Oliver Evans, now dead, had long wanted. In its first year it initiated a string of exhibitions that over time annually displayed the work of Pennsylvania entrepreneurs and mechanicians. At the end of two years the institute had more than a thousand members, was on a sound financial basis, and, as promised, was publishing a journal edited by a popular lecturer of the day, Thomas P. Jones, who roamed easily between the two worlds of mechanicians and entrepreneurs.

In his first issue, Jones announced that "the age of secrecy in arts and trades has nearly passed away.'' Consciously or not, he affirmed the attitude of Roswell Lee at the Springfield Armory and attacked the restrictive policy of his native land, England, but also reflected the views of the institute's founders, who were eager to promote at once the prosperity of the Commonwealth of Pennsylvania and that of the nation. They saw themselves, as Bruce Sinclair puts it, ''engaged in a large-scale cooperative effort on behalf of the Commonwealth and country. And individualistic, secretive technology smacked of old-world ways in their minds, and they sought to make the institute a focal point and clearing-house for information of all kinds.'' With that goal in mind, he published the latest patents, often with acerbic comments, in order ''to lay open those stores of the genius and skill of our countrymen which, although existing in the Patent Office, have hitherto been but very partially known.'' He bent every issue of the journal toward the practical—publishing a report from an institute committee on percussion caps for rifles, which it praised; a stream of articles on ways to increase iron manufacturing and the production of anthracite coal.

The education of mechanicians presented problems the institute strove to solve but never did. One wing of the leadership opted for a liberal-arts education, Latin and Greek required, which would open the door for entrance to the University of Pennsylvania. Others, like the remarkable Peter A. Browne, a lawyer who admired the mechanician as a man with ''an inventive, enlightened, and inquiring mind'' who needed the ''piercing ray'' of science but not Latin and Greek to promote his career, called for something close to what is known today as vocational education. Eventually, unable to resolve the split, the institute abandoned its educational program and turned it over to what became (and still is) the prestigious Central High School of Philadelphia. Browne's view had to await the imprimatur of Alfred North Whitehead in the twentieth century, who said that ''in estimating the importance of technical education we must rise above the exclusive association of learning with book-learning (that is, science)''; but in 1824, the founding year of the Franklin Institute, Amos Eaton of upstate New York also said that ''for the purpose of instructing persons . . . in the application of science to the common purposes of life,'' he had created a school to that end, which ultimately became Rensselaer Polytechnic Institute. The big leap forward came in 1862 when the Morrill Act called for federal support for schools that promoted the education of farmers and mechanicians.

The attitude of the institute and its journal—exemplified by the ridiculed notion "that artisans will, or can, become men of general science"—changed in 1829 with the arrival of Alexander Dallas Bache, who more than "any other single person" directed the institute into "experimental research, the professionalization of science and technology, and the elevation of America's scientific and technical reputation." He, with others of the new generation, transformed the *Journal of the Franklin Institute* "from a mechanics' magazine into a prominent scientific publication." Bache blended nicely a knowledge of science with practicality. He came to the institute at the age of twenty-three, a graduate of West Point with two years in the Topographical Engineers. As the great-grandson of Benjamin Franklin, he swam at once into a sphere of influence within the Franklin Institute and was immediately appointed to all key committees. He helped to broaden a once parochial outlook—what is good for Pennsylvania is good for the country—into a concern for issues that affected all parts of the nation, such as the best design for waterwheels and the cause of steamboat boiler explosions. Like Dr. Jones he believed mechanicians were "a class of our fellow-citizens whose importance we are only beginning to appreciate correctly," but how best to educate them was a question he could not answer.

Bache remained at the center of things in the institute until he left at the end of 1843 to head the U.S. Coast Survey, where he remained until his death in 1867. The institute had achieved much during its first twenty years but failed to answer the question that had preoccupied Bache. "By stimulating national pride in American manufactures, by providing the mechanicians themselves with news of the state of the mechanical arts and the science, by communicating in the pages of its *Journal* an awareness of the need for scientific method, the Institute facilitated the pace of mechanical progress," Anthony Wallace has written. "But the failure of the mechanics' schools, which were patronized more by the general public than by young mechanics, was perhaps an omen of an effect its own success was slowly helping to bring about: the

splitting of the mechanicians' fraternity into two, unequal, classes
—the class of the practical artisan-mechanic, with limited education; and the class of the engineer and the architect, who employed scientific knowledge to solve fundamental problems of
design.''

28

One More
Song-and-Dance Man

The tale goes that Benjamin Franklin and a friend saw the ascension of a balloon in Paris and the friend after the show remarked something like, "What good is that?" Franklin answered, "And what good is a newborn babe?" Franklin had created a new field in science—electricity—but his "babe" had yielded little that shaped daily lives except the lightning rod; the field had moved into the hands of pure scientists—natural philosophers, as they were called then. The first major breakthrough into the lives of the plain people came with the telegraph—a word from the Greek meaning "writing at a distance"—invented by Samuel Finley Breese Morse, a renowned artist of his time (and still honored). Morse does not really belong in a history of the nuts and bolts of the American past, but he was a fellow traveler, albeit a parasitic one. Few question that he marketed the first effective telegraph, yet historians, Brooke Hindle excepted—and even Hindle admits Morse was "mechanically inept"—tend to denigrate his achievement, largely because in later years he was less than generous toward those who guided him to fame and fortune. But Morse deserves space here because he illustrates a new trend in technology, the use of science and the findings of pure scientists like Joseph Henry to create technological innovations. With him, at last, the dictionary definition of technology quoted in the first chapter—"the application of science, espe-

190

cially to industrial or commercial objectives"—begins to become valid. Finally, Morse calls for attention because he epitomizes all the song-and-dance men who have done so much to advance American technology.

Morse was forty-one and the year 1832 when he returned from a European visit, already a distinguished painter but one who seldom had two nickels to rub against each other. Aboard ship a knowledgeable professor talked over dinner, and probably afterward as gentlemen passed the port among themselves, about electromagnetism and the recent surge of developments in the new field. Morse had some background in science, the one area in which he excelled during an undistinguished undergraduate career at Yale; he could comprehend much that he heard. Indeed, so the story goes, he said on disembarking, "Well, Captain, should you hear of the telegraph one of these days as the wonder of the world, remember that the discovery was made on board the *Sully*."

Morse had slipped into a scientific stream that was for years building to flood-proportions. The stream, then a trickle, begins in 1780 with Luigi Galvani, who, when Franklin still lived, noticed a twitching of frogs' legs hung by copper wires, when he touched them with a scalpel. An earlier generation might have assumed something supernatural had occurred; Galvani, an enlightened man, reported the oddity to the scientific world. Count Alessandro Volta, another Italian, followed up the puzzle and from his experiments uncovered a new form of electricity. Franklin had accumulated it through friction and stored it in Leyden jars, but with every experiment it gushed out, like water released by floodgates. Volta discovered an electricity created through chemistry—cloths soaked in brine layered with sheets of zinc and copper, which for some reason generated an electrical current that did not vanish with each experiment. His "battery," as it came to be called, opened a new world.

The next advance came from Denmark, where Hans Christian Oersted found by accident that a battery with the current running deflected a normally steady compass needle. A Frenchman, André-Marie Ampère, read Oersted's findings and quickly produced a paper "which laid down the entire basis of the science of electrodynamics." An Englishman, William Sturgeon, now moved on the scene and sent research down a side road. He found that by winding a bar of iron with wire attached to a battery he created a magnet. Joseph Henry, then teaching in an upstate New York academy, wound a bar of iron with insulated wire and produced a more powerful magnet. Innumerable people had by now sensed that clicks and clacks from one electromagnet to another could send messages over a distance, but all knew that the intensity of the electric current faded over distance. Henry solved this obstacle with what is today called an electric relay. "By using relays and batteries at regular intervals, there is nothing, in principle, to prevent one from sending a particular pattern of clicks around the world," Isaac Asimov writes. "By 1831 Henry was sending signals across a mile of wire." Henry left his achievement at that, for he "had no desire for fame and less for profit," Burlingame remarks. "His compelling motive was to increase knowledge of the natural forces and their application."

Burlingame for once errs. Henry wanted fame. He became terribly uneasy, for instance, at the possibility that Michael Faraday would precede him in the publication of some experiments. But fortune did not interest him. Morse learned about Henry's work from a colleague at New York University, where Morse was a professor of literature of the arts of design. He and Henry, now at Princeton, talked at length, but in old age Morse said he was "not indebted" to Henry "for any discovery in science bearing on the telegraph." He dismisses him as one whose work was "jackdaw dreams." Morse, sounding like Fulton a generation earlier (whose defensive letters he may have read), "readily admits that, in the construction of this telegraph, he uses many things invented by others," that "the chief merit, he claims, is that of so combining together things and inventions already existing, as to produce a result never before attained." It is at this point that Hindle comes forth to praise Morse, as earlier in his book *Emulation and Invention* he extolls another, less talented artist, Fulton. "The primary strength" Morse "brought to the telegraph was an excellent design capability based upon a mind practiced in forming and re-forming multiple elements into varying complexes. This sort of synthetic-spatial thinking is required in its most unalloyed form in painting and sculpture where analytic, logical, verbal, or arithemetic thinking plays almost no role."

Two items are missing from Hindle's judgment. First, Henry could have, if he had cared a damn, invented the telegraph as Morse conceived it; science rather than practical innovations intrigued him. The financial security that Morse hungered after Henry already had. Second, Hindle plays down the song-and-dance side of Morse, needed to sell his invention. He had perfected it at an inopportune time. A depression that began in 1837 and lasted into 1842 deterred investors in dubious ventures. Morse carried his agreeable personality to Washington, but it took until 1843, when the depression had waned, to entice from Congress what no one had ever been able to win from the federal government—a $30,000 grant to erect a telegraph line from Washington to Baltimore. Morse displayed his ineptitude in practical matters by wasting $23,000 of that subsidy attempting to bury his lines underground. Ezra Cornell from upstate New York saved him by stringing the lines aboveground with the pittance left, and in May 1844 the words "What hath God wrought?"—chosen from the Bible by a young lady, not by Morse—clicked along the wire. Morse might have preferred the message he had selected six years earlier for a private exhibition: "Attention universe, by kingdoms right wheel."

Morse expected his invention to supplant the mails; when the public and investors showed little interest—to promote business chess games were played over the wires—he tried to sell his creation to the government for the flat fee, patent included, of $100,000. By then the Mexican War had broken out, and thus by historical accident an improvident government left to private enterprise the development of what in Europe soon came under the domain of governments. It cost $185 or less to erect a mile of wire, and once newspapers and business saw what the telegraph could do for them, ribbons of wire spread everywhere east of the Mississippi. By 1852 the lines extended southward to New Orleans and westward to St. Louis, and coursed through more than five hundred towns and villages along the way.

The telegraph annihilated distance. Space, once measured in miles, was now measured in the moments between the time a man in New York pressed a key and another in New Orleans replied. The conduct of commerce was revolutionized. An English visitor reported in 1852:

If, on the arrival of an European mail at one of the northern
ports, the news from Europe report that the supply of cotton or
of corn is inadequate to meet the existing demand, almost before
the vessel can be moored intelligence is spread by the Electric
Telegraph, and the merchants and shippers of New Orleans are
busied in the preparation of freights, or the corn-factors of St.
Louis and Chicago, in the far west, are emptying their granaries
and forwarding their contents by rail or by canal to the Atlantic
ports.

By this time Morse had become a rich man. Meanwhile Joseph
Henry had moved from Princeton to head the Smithsonian Institution, which had been chartered by Congress in 1846.

29

Water and Science

America took slowly to the stationary steam engine as a source of power for machines, except in areas like that surrounding Pittsburgh, where the abundance of accessible soft coal offered a dependable and relatively inexpensive source of fuel. But in the East, where the lodes of anthracite in northeastern Pennsylvania were hard to reach and harder to move to market, water continued to be the principal source of industrial power. It was cheaper and cleaner than coal, self-renewing, and nonpolluting; but draw-backs—ice, floods, droughts, dams that silted up—made this gift from God an unreliable source of power, even in a region of swift-running streams. A shop or factory or mill usually could not depend on it to turn the waterwheel more than 160 days a year.

Waterwheels came in various shapes and sizes—overshot, undershot, pitchback, breast—and every mechanician and entrepreneur had an opinion, based on experience and intuition, about the one best suited to his stream, his needs. In 1829 a committee of the Franklin Institute, with the needs of the nation in mind, not just Pennsylvania, remarked that the efficiency of these various wheels "has never been fix'd by actual experiment" and appealed to owners around the country to support an investigation that would judge "the value of water as a moving power." The Middle States and New England gave spiritual and financial support to the project; the South expressed little interest. The institute inaugurated a series of precise experiments—to be exact, 123, which involved 1,381 trials extending over two years, each repeated at

195

least once—that analyzed every aspect of waterwheels and the amount of water needed to move them. The results were published in the journal and, as Wallace puts it, "created an international sensation," for they presented for the first time massive evidence based on rigidly controlled experiments on the efficiency of waterwheels under a variety of conditions. The research was a historic achievement, the first effort to apply scientific experimentation to a national problem; but it led nowhere because individual owners of waterwheels could find little in the published findings that seemed directly tied to their particular situations. Waterwheels continued, in size and shape, to turn as they had always turned, but between 1830 and 1840 Europe forced Americans to look anew at the use of waterpower to propel machinery.

One conclusion of the institute's experiments made it clear that the typical waterwheel, regardless of size or shape, was inefficient; that is, "the ratio of power expended to the result produced" hovered around 65 percent. This fact told entrepreneurs nothing new. For years, perhaps spurred on by Oliver Evans's *The Young Mill-Wright and Miller's Guide,* which Layton holds "did much to spread the scientific approach," the imaginative among them had been searching for ways to get more out of the water that flowed through their millraces. Early in the nineteenth century Benjamin Tyler of New Hampshire designed a wheel that had "the principal features of the modern turbine" but was abysmally inefficient. Later two brothers, Zebulon and Austin Parker of Ohio, modified the Tyler design; the Franklin Institute tested it in 1846 and found it "gave a maximum efficiency of 65 percent," which was 25 percent better than Tyler's but not enough to hasten a shift from waterwheel to turbine (a word rooted in Latin and meaning "spinning thing").

Meanwhile, Benoît Fourneyron, a French engineer, had created a turbine derived from mathematical calculations. The *Journal of the Franklin Institute* published a translation of his findings in 1842, and soon afterward a Philadelphia firm began to manufacture his turbine. It was expensive, but Eli Whitney, Jr., one of the first in southern New England, purchased one with the hope that it would diminish his growing problems with water power. The journal report intrigued Uriah A. Boyden, an engineer with a sound mathematical background, and he, too, bought a Fourney-ron turbine and had it sent to Lowell, where he lived and worked. He studied the machine, incorporated several modifications, and patented them. The Proprietors of Locks and Canals bought his patent rights in 1849, which brought Boyden in close contact with its chief engineer, James B. Francis.

Francis was an exemplar of the brain drain, even more impressive than Slater and certainly more imaginative. He had come from England at the age of eighteen and had found employment under George W. Whistler, then involved in building a railroad. When Whistler moved to Lowell—where his son, James Abbott McNeill Whistler, was born—he took young Francis with him. Soon after, in 1837, Whistler went to Russia to build a railroad for the czar; Francis stayed behind, having been chosen at the age of twenty-two to be chief engineer for the Proprietor of Locks and Canals, which made him responsible for maintaining all of the power that ran the now-numerous mills that flanked the Merrimack River. Francis saw that the turbine offered a possible way to improve efficiency of the power system and persuaded his superiors to subsidize experiments with the machine. Struik remarks that it "was one of the first instances on record in which a private American company appropriated a considerable sum of money for industrial research."

The partnership that developed between Boyden, a loner, and Francis, a sociable gentleman, was ideal. Francis was a superb mechanician and had the patience for lengthy and precise experiments. Boyden saw that the mathematical work of the French had flaws—it failed, for instance, to consider friction and internal resistance in theories about the flow of water. Together, as Layton remarks in an excellent essay, they "developed a scientific tradition that differed from the European one. Their bias against mathematical theory was not based on ignorance. Boyden, in particular, was an able and creative mathematician, albeit in an archaic, Newtonian geometrical tradition. Boyden, and probably Francis, were familiar with the mathematical theories of the turbine developed by Euler, Poncelet, and Weisbach. But they thought that these theories did not come close enough to reality to be of value to the practical man."

The turbine that resulted from this collaboration was called,

rightly, the Francis turbine. The experiments that led to its efficiency, approximately 90 percent, were those of Francis, who developed "turbine testing into a matter of exact science." It was
he who publicized the findings in his monumental *Lowell Hydraulic Experiments* in 1855; the book went through several editions and remained a reference work into the twentieth century.
Finally, he trained a number of assistants who carried his "scientific style of industrial research," which emphasized experimental investigation over mathematical theory, to all parts of the
country. One of these men, James B. Emerson, created a company
—the Holyoke Testing Flume—which served as a laboratory for
turbine research. The result of Emerson's tests "was a spectacu-
lar improvement in the performance of Francis turbines and an
even more remarkable decline in their cost. By the 1870s variants
of the Francis turbine became the most widely used hydraulic
prime movers in America, at a time when water power was still
more important than steam for industrial purposes."

There are hints of occasional strains in the partnership between
Boyden and Francis—a memorandum left behind by Francis im-
plies a fear Boyden was stealing an idea from him—but the
friendship remained intact. When Boyden died in 1879, his will
named Francis as a trustee to administer a fund of $230,000 to
build an observatory where there would be a minimum of "im-
pediments to accurate observations." According to Struik, that
"observatory was built by Harvard College, and stands at an
altitude of more than 8,000 feet near Arequipa in Peru."

VIII

The Future Rolls In

30

Wood

In 1853 a mixed commission of visiting English mechanicians and entrepreneurs deprecated America's ability to work metal into finished products but admired to the point of astonishment the technological progress in working wood. From the earliest settlements at least until the Civil War America flourished in an age of wood, long after Britain had moved into the iron age. The tools on a farm—rakes, plows, butter churns, hoes, washtubs, buckets—were made entirely or mostly of wood. The streets in towns were often paved with planks or blocks of wood. Bridges were built of wood. The looms in textile mills were mainly of wood. When the Pennsylvania oil fields opened, drillers stored their black gold in wooden tanks, then shipped it to the refinery in wooden barrels on flatcars built of wood.

"In no branch of manufacture does the application of labour-saving machinery produce by simple means more important results than the working of wood," one of the commissioners reported. He went on to speak of a shop visited that produced tapered shingles "at the rate of from 7,000 to 10,000 a day," of a machine that turned out 4,500 matches every minute, of a planer that sent out finished boards "at the rate of 50 feet per minute," and "at the same time that the face of the board is planed, it is tongued and grooved by cutters." The specialization he found in shops especially impressed him. "Many works in various towns are occupied exclusively in making doors, window frames, or staircases by means of self-acting machinery such as planing,

tenoning, mortising, and joining machines. In one of these manufactories twenty men were making panelled doors at the rate of
100 per day.'' The British visitor assumed that entrepreneurs
installed all of these machines to cut labor costs in a land supposedly faced with a constant shortage of workmen, especially skilled
workmen. In fact, Eugene Ferguson observes, drawing on recent
scholarship, the woodworking machines were ''designed not to
eliminate but to speed the work of skilled craftsmen.'' The entrepreneur still needed able mechanicians at the helm to ''guide,
direct, and take care of'' machines, which eliminated ''much of
the muscular work'' while ''the brain work remained.''

While English wheelwrights continued to build carts from raw
lumber to the finished product under one roof, Americans were
producing them en masse. The Studebaker brothers, wagon makers, bought their spokes, felloes, hubs, and other parts from shops
that specialized in those items and soon created a plant in South
Bend, Indiana, notable for its ''order, system, intelligent supervision, the best of material, and all the mechanical helps that
genius can contrive and capital produce.'' In the process they
reduced the price of their wagons from $140 to $70 with no loss
in quality. By the time of the Civil War, according to Alfred D.
Chandler, ''the nation's largest carriage-makers, using the most
sophisticated wood-cutting machinery, the minutest subdivision
of labor, the most carefully designed plants, and nationwide marketing agencies, had an output of 40,000 to 50,000 carriages a
year.''

Mass production led inevitably to a standardized product. A
New Haven carriage-maker raised its production from 3 a week
to between 2,500 and 3,000 a year in part because it concentrated
on a single style. This drive for a standard product at least in
housing, for which there was a constant, desperate need, had deep
roots in the American past. Philadelphia carpenters as early as
1724 could buy ready-made sash windows for houses. In 1795
another local carpenter advertised lumber ''of good quality and
well seasoned'' ready for houses twenty feet wide, forty feet deep.
The need for a house that could be raised quickly became an
obsession by the 1830s as a new wave of emigrants began to flood
into the country. Augustine D. Taylor of Chicago is now credited
with fulfilling the need in 1833 with one of the most radical in-

Balloon-Frame House. Despite the contemptuous nickname old-time carpenters gave the seemingly fragile structure, it turned out to be as sturdy as the traditional house built of posts and beams sixteen inches square. An early historian hailed the innovation in building as ''the most important contribution to our domestic architecture which the spirit of economy, and a scientific adaptation of means to ends, have given the modern world.'' *(Metropolitan Museum of Art, Harris Brisbane Dick Fund, 1934)*

novations in the history of construction—the balloon-frame house, so called for its fragile appearance. ''Basket-frame'' is an apter name, as Daniel Boorstin makes clear in his description of the invention :

The basic new idea was so simple, and today is so universally employed, that it is hard to realize it ever had to be invented. It was nothing more than the substitution of a light frame of two-by-fours held together by nails in place of the old foot-square beams joined by mortise, tenon, and pegs. The wall plates (horizontals) and studs (uprights), the floor joists and roof rafters were all made of thin sawed timbers, . . . nailed together in such a way that every strain went in the direction of the fibre of the wood (i.e. against the grain). "Basket-frame" was the name sometimes given it, because the light timbers formed a simple basketlike cage to which any desired material could be applied inside and out. Usually, light boards or clapboards covered the outside. Nothing could be simpler. Today about three-quarters of the houses in the United States are built this way.

The houses were "often appallingly unattractive," but as Solon Robinson, a New England migrant to the Midwest in 1834, said, it was "the most important contribution to our domestic architecture which the spirit of economy, and the scientific adaptation of means to ends, have given the modern world."

Taylor's invention depended on another that had come nearly forty years earlier—Jacob Perkins's nail-making machine of 1795, which was said to have produced 200,000 nails a day. Since then scores of mechanicians had improved on his creation, and the price of nails had dropped precipitously, from about twenty-five cents a pound in 1795 to three cents by 1842. By that time, according to Robert Fogel, "the domestic production of nails probably exceeded that of rails by over 100 percent." As the cost of nails fell and news of Taylor's basket-frame house spread, America's consumption of wood soared to more than 1.5 billion board feet a year, or approximately 100 feet per person. The timber came from 31,000 sawmills scattered about the country, some 6,000 in New York alone, another 5,000 in Pennsylvania.

Technological advance came slowly in the sawmills, in part because the owners, with an abundance of wood at hand, did not need to economize but also because America's iron and steel industries could not give them better blades. The waste was appalling to outsiders. "Lumber manufacture, from the log to the finished state, is, in America, characterized by waste that can truly be called criminal," said a British visitor. Nathan Rosenberg reports that of 1,000 feet of board that went through a circular saw, 312 feet ended up as sawdust. "If the saw could be reduced to one-twelfth of an inch, which was the thickness of the early band saw, only 83 feet would be lost in sawdust." When the band saw and later the jigsaw came in, not only did the mills become more economical in the production of lumber, but the innovations in technology had a startling effect on house design. Taylor's houses were stark, unadorned structures, "bare, bald white cubes," an architect called them, which reflected Americans' lack of "cultivation of the higher natural perceptions." The band saw was "ideally suited to carving curved forms"; the curlicues and scrolls that soon began to decorate the exterior of houses were "artistic" touches that delighted architects and their clients. They gave a touch of class to a standardized product.

Mass production of standardized products emerged from America's woodworking industry, along with a host of mechanicians and machinery adaptable to the working of metal once the country acquired an advanced technology in that area, borrowed or stolen largely from England. Steamboats east and west created a pool of mechanicians throughout the North that could shift to railroad shops when the time called for it. Planers that smoothed wood could be made to do the same for metal. Lathes and saws that cut and ground metal arose from machines the workers in wood had created. Two streams of American technology—one flowing from the government armories that worked mainly in metal, the other from the manufacturers who worked in wood—merged in the 1840s as the use of anthracite coal spread and America belatedly entered the iron age.

31

Coal

Down to the 1830s Philadelphia powered its few steam engines with wood or with bituminous coal imported mainly from Newfoundland or England and commonly known as Liverpool coal. Everyone knew that less than a hundred miles north of the city lay vast deposits of anthracite coal; indeed, as it turned out, they were the largest deposits in the world. Tench Coxe, whose family fortune was to be built on his prescience and on the claims he staked out in northeastern Pennsylvania, had said since the early 1790s that "when our wood-fuel shall become scarce, and the European methods of boring shall be skilfully pursued," Pennsylvania would prosper mightily. Few at the time listened to him. The coal lay in mountainous retreats then largely inaccessible. Also, "stone coal," as it was called, was hard to ignite and once aflame demanded steady currents of air to keep the bed burning, something stoves and furnaces of the day were not equipped to provide. During the War of 1812, when Philadelphia was shut off from Liverpool coal, two entrepreneurs tried to sell a shipment of stone coal to the city—without luck.

Then technology stepped in. In 1819 Jordan Mott of New York City, which had long had to import its wood for heating and cooking, much of it from New Jersey across the Hudson River, invented a stove that would burn small pieces (nut coal) of anthracite. He began to manufacture his stove, and its success caught the interest of entrepreneurs eager to milk the great lode that lay in Pennsylvania. They hired John Jervis, an ex–farm

boy from upstate New York who had proved talented while helping to construct the Erie Canal, to create another canal across
New Jersey and a railroad or whatever—no one knew what—up
the western side of the Delaware River into the Pennsylvania
mountains. Jervis did so in the late 1820s and opened a once-
isolated region to exploitation. Philadelphia entrepreneurs, belatedly, soon followed with canals and later railroads of their own
that directed the flow of coal southward to their city. The Franklin Institute, as we have seen, stepped in with the offer of prizes
for the best anthracite-burning stoves and with exhibitions that
publicized the beauties of anthracite coal. A new age was
launched. This account condenses a complicated story almost to
an absurdity, but it seeks to make the point that in the late 1820s
a vast natural resource long ignored became used as needs (shortage and rising cost of wood) joined hands with the entrepreneurs
and advances in technology.

The use of anthracite proliferated like rabbits. It was cheaper
and cleaner than soft coal; a single bushel gave as much heat with
less soot and smoke as three bushels of soft coal and, according to
Chandler, "there is a saving of about fifty percent." It also was
cheaper, burned longer, and demanded less tending than wood.
The head of the Pennsylvania Hospital found in 1825 that his
cost of heating and cooking dropped from $3,200 to $2,100 when
he switched from wood to anthracite. By the mid-1820s the coal-
burning cast-iron stove, warmly endorsed by a stream of articles
in the Franklin Institute's journal, had become a standard item
in most northern homes. The cost ranged from five to twenty-five
dollars. Mott sold his with a supply of nut coal to entice buyers,
"and it was not long before stoves were being marketed with all
the hoopla that we have come to expect in the sale of consumer
durables: large illustrated advertisements in newspapers and
magazines, extravagant claims, competitions at county fairs,
installment-buying plans, attractive giveaways (such as extra kettles and frying pans), licensed traveling salesmen, and 'free' do-
nations to prominent citizens." Aesthetes, with "a feeling of
unutterable repugnance," mourned the displacement of the fireplace in the home. It "will never cease to be loved for the beautiful atmosphere it imparts to a room," said one, but a woman who
had a house to run said, "You may take the poetry of an open

wood fire of the present day, but to me in those early days it was only dismal prose, and I am grateful to have lived in the time of anthracite coal.''

New uses for coal were found as the Pennsylvania fields became more accessible through the network of canals and railroads; output rose from 210,000 tons in 1830 to 1,164,000 in 1837. Solomon Willard, ''a talented jack-of-many-trades'' in Boorstin's words, invented a hot-air furnace that fed heat by pipes throughout a house. Eliphalet Nott, a clergyman and president of Union College, designed a boiler fed by anthracite that in 1836 carrried a steamboat on an experimental trip from New York to Albany on twenty tons of coal at a cost of five dollars a ton. ''The same voyage would have consumed forty cords of pine wood,'' said a passenger, ''the present price of which is six dollars, making a difference of more than half.'' Railroad locomotives, which began by burning wood, soon shifted to coal. The rise of gas illumination further increased the use of anthracite in eastern cities. England had pioneered the new way to light the night, but America caught up quickly once coal came in. Rembrandt Peale, son of Charles

Loading anthracite coal onto a barge in the Lehigh Canal. *(New York Public Library)*

A coal barge. *(New York Public Library)*

Willson Peale and the mechanician much admired by young George Escol Sellers, used gas to illuminate an adjunct of his father's Philadelphia Museum in Baltimore in 1816, and soon a gas company was chartered there. New York City created one of its own in 1823. A decade later Philadelphia sent Samuel V. Merrick, one of the founders of the Franklin Institute, to England to study gasworks there. The creation of a Philadelphia gasworks, which came about soon after his return, was promoted mainly by entrepreneurs who had holdings in the anthracite region.

Once the mine operators learned in the early 1840s that anthracite rather than charcoal or coke could be used directly in making iron, production in the Pennsylvania field soared from 1,900,000 tons in 1844 to 3,327,000 in 1847. "Thereafter," Wallace reports, "in the anthracite area, one innovation followed another in the rush to increase production. In 1844, the first of the new steam-powered mechanical coal breakers invented by Joseph Battin was put in operation; thereafter the huge wooden breaker house where the lumps of coal from the mines were broken and sorted into sizes began to dominate the landscape at the mines. New elevators, new exhaust fans, new Cornish bull steam engines, new diamond drills, new water pumps came on in an endless parade of mechanical ingenuity." Safety features belowground did not keep pace. Years earlier English miners had learned of the danger from methane, an odorless gas that accumulated at the rockface where a miner worked; the flame from his candle often ignited it into a terrifying explosion. Sir Humphry Davy and his assistant Michael Faraday designed a safety lamp early in the nineteenth century. Paradoxically, it promoted explosions, for miners, unhappy with the weak light it gave, often resorted to candles. After a long string of further disasters, Faraday, at the government's request, restudied the problem of methane in 1844 and concluded that ventilation shafts would carry the deadly gas away from the workman's face. American operators knew of his findings, because the Franklin Institute's journal printed an abstract of his paper the following year, but they paid little heed to it. Such shafts took time to sink and cost much money in a business in which profit was touch and go. "It was the ventilation system that was regularly disregarded by operators who did not bother to make use even of the technical knowledge readily available to them in English and American treatises on coal mining methods," Wallace says. As far as the mining of Pennsylvania coal went, he adds with some acerbity, "the Industrial Revolution would appear to be a game played by technological gamblers who liked to bet their own and others' lives and money against disaster.... Do we see here the prelude to environmental disasters of the twentieth century?"

32

Iron

The way Americans made iron changed little from the foundation of the first settlements until the early 1800s, and the few innovations adopted over time came from Great Britain. The iron plantation, a typical production unit, varied in size from five thousand to ten thousand acres. It was an isolated, self-contained settlement of some half-hundred men and their families with an equal number of draft animals that mined a local deposit of ore and used a surrounding forest for fuel to smelt the ore into pig iron, so called because the line of sand molds into which the ore was poured resembled a line of piglets suckling from a sow. The refining process was tedious. The ore was dumped between layers of charcoal and lime—in the East oyster shells served as a flux to draw off impurities—into a brick cylinder or stack about thirty feet high. Bellows powered by a water wheel gave a steady blast of air to the fire. A typical plantation produced about a thousand tons of pig iron a year, which went to foundries that cast the pigs into kettles, frying pans, and other cast-iron items, or to forges that hammered out the remaining impurities and created a malleable or wrought iron.

The innovation of coke—soft coal from which impurities had been extracted, leaving behind a fuel of concentrated carbon that gave off an intense heat—came from Britain. It meant little to plantations in the East, which lacked a cheap and abundant source of bituminous coal; they stuck to charcoal while entrepreneurs in the Pittsburgh area began to shift to coke. Next, the

British created the reverberatory furnace. Here, in a box within a box, as Elting Morison succinctly puts it, "the pig iron was melted down from the heat of a fire that was separated from the iron by the interior partitions of the furnace. The metal in a molten state was then stirred or puddled by heavy wands or rods inserted through ports in the furnace wall." Slag, drawn to the top by lime, was scooped off, and wrought iron flowed from the bottom of the stack. But the whole process remained a hit-or-miss affair. Some batches came out brittle and useless for converting to refined products; others were perfect. No one, since all were ignorant of the chemistry involved, knew why. Even the most experienced puddler often had to step back in bewilderment at a batch that had failed. Burlingame likens "the whole performance" to "a cook tasting as she stirs, to make sure there is enough seasoning or, perhaps, of a painter daubing his canvas, standing back to look at it and then, with his thumb, removing some of the paint."

America as late as 1830 produced only 200,000 tons of pig iron annually, nowhere nearly enough to satisfy the expanding nation's needs. Britain, with a century's head start, at least to 1850 kept America in bondage for finished iron and steel products. Most American mechanicians who designed tools for themselves and their shops sent prototypes to England for reproduction. Excerpts from manifests collected by Robert Albion from ships entering the port of New York around 1850 suggest the degree of dependence of America on Britain for finished metal goods. Among the listed were compasses, thimbles, hammers, gimlets, wire, knitting needles, shovels, skates, pokers, locks, dustpans, gridirons, teakettles, carpenters' rules, chains, spittoons, tongs, dog collars, surveyors' chains, files, anvils, spectacles, frying pans, and buttons.

Despite the lag in America in producing the fine finished products English mechanicians, from long experience, excelled in, the sudden accessibility of anthracite planted seeds that grew to shake the iron industry from its colonial past. In the East anthracite quickly became cheaper than wood, charcoal, soft coal, coke, and even imported Liverpool coal; and low transportation costs, coupled with the rise of the railroads and the opening of new lodes in central and western Pennsylvania, ended the long reign of iron plantations. The refinement of ore moved from isolated regions to urban centers close to the anthracite mines—Bethlehem, Lehigh, Allentown, Norristown, to name a few in Pennsylvania. Coal operators joined with iron manufacturers to find efficient ways to use the new coal to make iron more cheaply than the British. Their greatest achievement was reproducing in the iron industry what Lowell had achieved in textiles—the integrated factory.

Morison has singled out John Fritz as an exemplar of this new trend in the iron industry. Fritz was reared on a farm in southeastern Pennsylvania. In 1838 at the age of sixteen he was apprenticed to a local blacksmith, who taught him to make horseshoes, iron rims for wagon wheels, and parts for gristmills and sawmills. When twenty-two he was put in charge of an iron plant at Norristown, and from there he moved to a site along the Susquehanna River where he created something revolutionary, an integrated mill—"the slow putting together of all the elements from smelting through all the states of refining and finishing. With a crew of Pennsylvania Dutch boys just off the farm, he built the plant from the ground up. Moving the heavy parts of the machinery and equipment by hand and putting them together with 'two-handed chisels and sledges,' they assembled the entire plant—mill and blast furnace—in about a year." Fritz went on to fame, at least in the iron industry, in Johnstown, Pennsylvania, but as he looked back in 1890 to the new world he had done so much to create, he deplored the view. The mechanician he had once known seemed to have vanished. "Present shop practice," he said, "is better calculated to make machines out of men than to make good all-around mechanics."

33

The Old Order
Begins to Pass

Jefferson's hope that household manufacturers and machine shops could successfully compete against England's industrial might had ebbed by the time he died in 1826. "By the early 1850s," with the rapid and prodigious spread of anthracite coal throughout the Northeast, Chandler remarks in a statement hard to improve upon, "American manufacturing had begun to move out of the shop and mill and into the factory. Its setting had become urban rather than rural." The appearance of a cheap, relatively clean fuel brought the stationary steam engine, before this time a rarity, into its own. Water power did not encourage a concentration of industry, except at choice spots like Lowell and Lawrence, but steam engines driven by cheap coal drew machines from the garden into cities. "Large steam driven factories were unaffected by summer drought and spring freshets," Chandler goes on. "The workers in these new factories were no longer farm girls, but men, heads of families, who had moved permanently to the city from the countryside of the United States and increasingly from the farms of Ireland, England, and Germany. These men no longer had any connection with the land. For them, the factory wages had become the sole source of income."

The changes would not have demolished Jefferson's faith in America's future, but they would surely have disheartened him as much as they did John Fritz, born nearly a century after him.

"Tasks in the factories were far more routinized and subdivided than in the shops and small mills," Chandler's summary of the passing of the old order continues. "The relations with the owners had become distant and totally impersonal. The workers in these factories were, by the 1850s, as they had not been in the 1830s, members of a proletariat in the Marxist sense." The railroad "universalized" the abrupt change. "The rushing locomotives brought noise, smoke, grit, into the hearts of the towns," Lewis Mumford writes. "And the factories that grew up alongside the railroad sidings mirrored the slatternly environment of the railroad itself." Sentimental historians choose to forget what Mumford knows, that American cities before the railroad were not antiseptic havens: they were filthy, they reeked of privies that needed emptying, of dead animals rats gnawed to skeletons before anyone bothered to haul them away. The point to emphasize is not the deterioration of urban life but the *change* that coal had brought to it.

Years before the East sensed the old order was passing, Michel Chevalier had seen the wave of the future in Pittsburgh in 1838. "It is," he said, "surrounded with a dense, black smoke, which, bursting forth in volumes from the foundries, forges, glasshouses, and the chimneys of all the manufactories and houses, falls in flakes of soot upon the dwellings and persons of the inhabitants; it is, therefore, the dirtiest city in the United States." Chevalier did not disparage the city, even though he did not like it. He saw that the Midwest needed what Pittsburgh made— "axes to fell the primitive forests, saws to convert the trees into boards, plough-shares and spades to turn up the soil once cleared. It requires steam-engines for the fleet of steamers, which throng the western waters. It must have nails, hinges, latches, and other kinds of hardware for houses; it must have white lead to paint them, have furniture and bed linen, for here every one makes himself comfortable."

The frenzy of activity in Pittsburgh—"there is no interruption of business for six days in the week, except during the three meals, the longest of which occupies hardly ten minutes"—barely equals what he found in the anthracite region. The frontier Frederick Jackson Turner found vanishing in the West in 1893, Chevalier had found alive in eastern Pennsylvania nearly half a century earlier. He visited a desolate spot already named Port Carbon. It held in 1838

> about thirty houses standing on a declivity of a valley, and disposed according to the plan of the embryo city. Such was the haste in which the houses were built, that there was no time to remove the stumps of the trees that covered the spot; the standing trees were partially burnt and then felled with the axe, and their long, charred trunks still cumber the ground. Some of them have been converted into piles for supporting the railroads that bring down the coal to the boats; the blackened stumps, four or five feet high, are still standing, and you make your way from one house to another by leaping over the prostrate trunks and winding round the standing stumps.

The seldom-quoted Chevalier liked what he saw in America as the old order was passing away. True, the country had "many imperfections, . . . But a few errors and follies are of little import in the eyes of those whose thoughts are occupied with the great interests of the future rather than with the paltry troubles of the present hour." The country does not tolerate "an idle tourist, seeking only for amusement!" Let him smile sardonically at the American's "simplicity and extravagance of national vanity. That patriotic pride, rendered excusable by brilliant success, will be moderated; the errors and follies are daily correcting themselves; the unavoidable rudeness of the backwoodsman will be softened, as soon as there are no more forests to fell, no more swamps to drain, no more wild beasts to destroy. The evil will pass away, and is passing away; the good remains and grows and spreads, like a grain of mustard."

IX

"The Great, Straddling, Bellowing Railway"

34

The Challenges Presented

With the railroad, technology and tradition joined in combat early in the 1830s. Historically, chartered roads, like canals, had to receive all common carriers who paid the required toll. Pennsylvania's state-financed Columbia Railroad expected when it opened to move only privately owned or leased carts pulled by horses along the line, and one of its designers told a doubting Quaker farmer exactly that. "My good Friend," he said, "I would have you understand that this railway is not being made for steam power; it is a State road for the benefit of every one, just as any turn-pike road. The State may furnish wagons or carts or individuals may put their own on the road, and every farmer may attach his own horses and haul his produce to market, and if I have my way no steam engine shall ever run on the road."

Inevitably, even before the Columbia Railroad became part of the privately owned Pennsylvania Railroad, the steam engine came in, and with it freight cars and carriages owned by the road. Soon the courts redefined the ancient phrase "common carrier" to include the railroads. Here Herman Melville's father-in-law, Lemuel Shaw, the chief justice of the Massachusetts Supreme Court, led the way in adapting common law "to the needs of railroad enterprise." He coined the phrase "eminent domain," an "American legal invention" that "enabled the state to buy private property for a public use, which now included the use of railroads," in the words of Boorstin. "Shaw thus recognized that, although the railroad company ran only its own cars on the line, it was actually a new kind of public highway."

The English experience, nearly a generation ahead of America's, gave less guidance than Americans hoped it would. The English had built their roads and equipment to last until the Second Coming. With short distances to cover, they could afford to be meticulous. They seated iron rails on blocks of granite. Their ponderous locomotives had sturdy iron bridges to hold their weight while crossing streams; they were underpowered by American standards but had few steep grades to face in the rolling countryside. Wood-rich, iron-poor (at least to the 1850s), America began with wooden rails topped by strips of iron and bridges made of wood; they called for lighter locomotives, yet powerful enough to curve their way up the sharp grades that prevailed in much of the country east of the Mississippi. The English arranged passenger cars with first-, second-, and third-class seats, a plan that, as it soon became apparent, would not do in the land of equality. The government subsidized construction of the English railway system. American state governments started to do the same—here Pennsylvania led the way—but the onset of the depression of 1837–1842 forced them to cancel their commitments. If America were to have railroads, entrepreneurs must do the job with a minimum of help from government. The obstacles they faced were formidable.

The foremost was money. The outlay, prior to the Civil War nearly all drawn from private sources, was astronomical for the times. No scholar has disputed Louis Hunter's estimate that "the construction of a single mile of a well-built railroad was enough to pay for a new and fully equipped steamboat of average size." And once that mile had been built there were, in many cases, taxes to be paid on the right of way. And unlike the steamboat owner who depended on the government to keep the rivers open, it had to be maintained, which involved the replacement of wooden ties, which America, with much space to cover and cost in mind, had substituted for granite block, new spikes for those broken, and new rails for those worn; and the gravel roadbed had to be periodically raked, shifted, and cleaned. By 1850 more than $370 million had been put into American railroads, a considerable amount of it drawn from English investors, and seven years later the total investment had reached a billion dollars.

Costs drove the entrepreneurs to develop short lines that connected areas of high potential traffic; thus a few miles here, a few miles there, did the roads push their way across the country. One of the first and most successful of these short lines was the Camden-Amboy, which created a lucrative link between New York City and Philadelphia without the expense of bridging rivers. Track that covered the sixty miles cost $3.2 million, half a million of which went for imported iron rails, another half-million to the roadbed. Locomotives, shops, carriages, and wharves added several hundred thousand more dollars to the investment. Despite all this expense, the road cleared close to $225,000 the first full year of operation, which helps to explain why American businessmen took so enthusiastically to the railroad.

Investment costs aside, the development of American railroads bears a close resemblance to that of steamboats. David Stevenson, an eminent developer of the British rail system, remarked on a visit to America in the 1830s, ''There are hardly two railways in the United States that are made exactly the same way, and few of them are constructed throughout their whole extent on the same principles.'' (His comments on American steamboats are similar: ''on minutely examining the most approved American steamers, I found it difficult to trace any *general* principles which seem to have served as guides for their construction. Every American steam-boat builder holds opinions of his own, which are generally founded, not on theoretical principles, but on deductions drawn from a close examination of the practical effects of different steamboats, and these opinions never fail to influence, in a greater or less degree, the built [*sic*] of his vessel, and the proportions which her several parts are made to bear to each other.'') The track gauge—the distance between rails—was standard throughout Britain, eight feet four and a half inches, the accepted width of an English cart or wagon axle. Once America began to make its own equipment, the short lines' gauges varied from region to region. New England accepted the English measure; the South chose five feet. The Erie in New York preferred six feet to prevent the diversion of traffic to Commodore Vanderbilt's competitive line in that state. By the time of the Civil War, says George Rogers Taylor, "there were still at least eleven different gauges in the North, and the 5-foot gauge was by no means universal in the South. Between Philadelphia and Charleston there were at least eight changes in width of track."

As the railroads spread over the land, they drew mechanical talent from all parts of the country into their orbit. John A. Roebling, a recent immigrant from Germany, in 1839 designed twisted wire rope to haul equipment on inclined planes up and over the Allegheny Mountains of Pennsylvania; he went on to build suspension bridges, one across the Niagara River that carried trains as well as people, and ultimately a span over the East River that survives today as the Brooklyn Bridge. Other less gifted mechanicians—boilermakers, pipe fitters, sheet-metal workers, car repairmen—drifted away from shops that made or repaired steamboats or built locomotives into railroad shops.

35

The Locomotive

The age of the moguls, when men made titanic fortunes from railroads, lay a generation ahead. That age's cast of characters— Jay Gould, Commodore Vanderbilt, James Hill, and the rest— has been remembered and, of course, disparaged with the words "robber barons." But the men who created the railroads from the ground up have suffered a worse fate: historical amnesia. Who, except scholars and railroad buffs, ever heard of John Jervis, Phineas Davis, Charles Danforth, or Joseph Harrison and "hundreds of lesser-known men who helped raise the impressive giants of the rails to their zenith?" Eugene Ferguson queries in a brief eulogy to one of them, John Brandt. "The boldness and audacity of these men, tempered by a finely intuitive sense of fitness, are aspects of the American character that we would do well to hold onto." Once again mostly unheralded entrepreneurs and mechanicians joined to solve the multitude of challenges the railroad presented to the country. Of all those challenges the locomotive called for the most and immediate attention. John Jervis, who has been recalled from the past in a memorable essay by Elting Morison, seems to have been the first to see that the sturdy English engine had to be adapted to American needs and conditions. Let a warning from Daniel Calhoun serve to introduce Jervis and all those involved in creating an American locomotive. "The men who did commonly design and build locomotives in this period," Calhoun writes in his history of American engineers, "were not engineers, at least not so titled, but were skilled me-

chanics who worked for manufacturers or became manufacturers
themselves. They were men like the 'Ingenious mechanic' Matthias Baldwin, who built up a large works in Philadelphia, or the
'ingenious Mechanic' Ross Winans of Baltimore, one of a series
of men who built engines primarily for the Batimore & Ohio.''

The lives of few Americans have spanned a time of greater
change than that of John Jervis, who was born in 1795 in upstate
New York, where the countryside had changed little since the
1600s; he died in 1885, two years after the Brooklyn Bridge
opened for traffic. He began life off the farm with an ax, leveling
trees to clear a path for the Erie Canal. He caught the eye of the
chief engineer, Benjamin Wright, and that led to building the
Delaware & Hudson canal, conceived to haul coal out of the Pennsylvania anthracite region to New York City. On that job he
experimented with steam engines to surmount by inclined planes
and a locomotive a 900-foot cliff at the western end of the 105-
mile canal. His next assignment led in 1830 to building the Mohawk & Hudson Railroad from Albany to Schenectady, at a time
that, says Morison, ''no one in this country knew with confidence
how to build a roadbed, what to use for ties and rails, or what
weight a locomotive could pull.'' Jervis ordered a locomotive
from England but found it too heavy for his hemlock rails, too
rigid to make the curves the terrain forced him to make. After
four years of experiments he produced a prototype with two driving wheels that held the weight of the boiler and engine and a
''bogie'' truck with four wheels that swung with the curves. He
thereby ''solved what had become the major problem in American
engine building and use''—how to keep the locomotive from
being derailed by sharp curves—and his ''scheme became standard American design for a long time.'' He went on to build the
Croton Aqueduct (still standing), which brought drinking water
into New York City in 1842, then he went back to railroad building—the Michigan Southern, the Chicago & Rock Island, the Fort
Wayne & Chicago (later the Nickel Plate). By the time he had
retired, the locomotive had evolved from his basic design into a
machine adapted to American needs—''a crazy affair, as loose-
jointed as a basket,'' said an English visitor. ''Let the road follow
its own wayward will, be low here and high there . . . the basketlike flexibility of the frame and its supports . . . adjusts the engine
to its road at every instant of its journey.'' Morison's words can

serve as his epitaph: "Jervis had within himself the sense of how to find the simple solution, which is probably one reason why he so often called the work he was doing an art."

In 1838, only a few years after Jervis had his prototype in running order, the federal government reported 337 locomotives chugging about the nation, all but 82 of them built in the United States. The figures defy belief. Whence came the resources, the talent to promote this industrial explosion? The exploitation of anthracite coal, as Chandler has shown, prodded the iron industry into the modern age, though it still lagged behind Britain (most of America's iron rails continued to come from overseas until the Civil War). The talent came, as in the past, from ex–farm boys turned mechanicians and small-time entrepreneurs inspired by the challenge of a new mechanical phenomenon. Phineas Davis, to cite a single example, was an ingenious blacksmith from the back-country town of York, Pennsylvania; local citizens called him a "prodigy in mechanics." He visited Peale's Museum in Philadelphia about 1828 and noticed on a backroom shelf the model of a

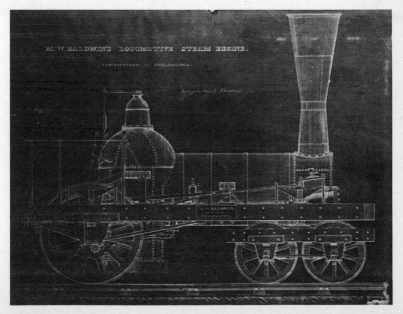

The Baldwin Locomotive, 1834 to about 1840. *(Smithsonian Institution)*

steam tractor that Oliver Evans and Charles Willson Peale had designed years ago for farm work. He tinkered until he got it going, and then, fascinated, ran it up and down the brick-walled yard behind the State House. Later in his visit he chatted with Matthias Baldwin, who had yet to build his first locomotive but had built a unique stationary engine. "I believe that the conversation with Baldwin, on the value of vertical cylinders, and the day with the Evans and Peale model gave the bent to Davis' mind," George Escol Sellers recalled in his memoirs. Davis went on to build locomotives for the Baltimore & Ohio and to run one of its shops. He died a few years later at the age of thirty-five in a derailment of one of his engines. "By this one accident," said Sellers, "the world lost a mechanic who had proved himself to have been in advance of the age, and whose name should be remembered among the most original men of the time, and a man who, had he lived, would not have been left behind in the race."

The locomotive builders came from a variety of backgrounds. Thomas Rogers of Paterson, New Jersey, had made a comfortable living manufacturing machinery for textile mills until John Jervis urged him "to try his hand at locomotive building"; then he made a fortune and soon became one of the most innovative manufacturers in the business. Matthias Baldwin began adult life as a jeweler and turned to making printing machinery for textile companies before becoming the preeminent builder of locomotives in the country. William Norris had been a Baltimore merchant until he teamed up with Major Long to build locomotives. George Whistler, a West Point graduate, left the army to build railroads

The English locomotive *John Bull*, 1831, was inside-connected with full cranks. *(Smithsonian Institution)*

in New England with his friend William McNeill (after whom he named his son), then concentrated on locomotives after he went to the shops at Lawrence; from there he went to Russia to build a railroad for the czar. Joseph Harrison, Jr., began as an apprentice mechanician in Philadelphia but as a young man started his own company. One of his locomotives won fame when it pulled a train of 101 loaded coal cars. He later joined Whistler in Russia and completed the contract with the czar after Whistler's death in 1849. Back in Philadelphia he swung to a new specialty and designed a steam boiler joined by bolts rather than rivets. Its virtue, he said, was that "it is easily transported and may be taken apart so that no piece need weigh more than eighty pounds. In difficult places of access, the largest boiler may be put through an opening one foot square."

One of the most remarkable of these men, an exemplar of versatility, was Stephen Long. After a tour of duty as an army engineer on the Mississippi, he explored the Rocky Mountain region—Long Peak is named after him—and the source of the Minnesota River. A new career began when he laid out the route for the embryonic Baltimore & Ohio. (One of his assistants at the time was Benjamin Latrobe, son of the builder of the Philadelphia Waterworks.) The partnership with Norris was short-lived, but the locomotive he designed for him eventually proved durable. Next, from 1837 to 1844, he served as chief engineer of the Western & Atlantic Railroad in Georgia. That experience marked him deeply. George Escol Sellers visited him early in the Civil War in Washington, where Long headed the Topographical Department for the army. He was appalled how rapidly the colonel, then seventy-eight, had aged since a previous visit two years earlier. Not the workload, heavy as it was, Long explained, had worn him down, but his many years in the South. "He had a warm friendship for some of those then engaged in 'this fratricidal war,' and it was the horror of the thought of that, and not the work that was killing him." He died two years later, in 1864, his life "shortened by the cares and anxieties of the first years of the war," Sellers thought. "I have always considered him the leading engineer of his time, and I am not alone in that, for B. H. Latrobe referred to him as the father of the engineers of his day."

36

New Founding Fathers

Herman Haupt, twenty-eight years old and a graduate of West Point, applied for a job in 1846 with the recently chartered Pennsylvania Railroad. Samuel V. Merrick, the president, told him in effect to get lost, that ''engineers were as plenty as blackberries.'' Haupt forever remembered Merrick as ''haughty and supercilious''; as the son of ''impoverished parents,'' according to his biographer, acerbic opinions of the elite came easily to him. Merrick, at the age of forty-five, was counted among the most eminent of Philadelphia entrepreneurs, a founder of the Franklin Institute, proprietor of a profitable plant that built fire engines and later parts for locomotives, a leader in introducing gas illumination into the city. But the new age that was dominated by the railroad did not suit his talents. Railroading called for another type of leader, and Merrick was soon supplanted by the road's chief engineer, J. Edgar Thomson, one of the most remarkable men of the time. As new spokesmen for America were arising in literature—Hawthorne, Emerson, Thoreau, Whitman, Melville—something similar was happening in the less noticed world of entrepreneurs. Leaders of the old guard were being replaced by gentlemen like Whistler, Latrobe, Haupt, and Thomson, who were intimately familiar with the mechanicians' world and, like them, salaried employees. They called themselves engineers, had much experience in building and operating railroads, and were committed to a lifetime career in their newly created profession.

Thomson had come to the Pennsylvania Railroad with more

than fifteen years of experience building railroads in the South. When he offered Haupt a post with the road, Haupt misjudged the short, taciturn man until a mutual friend said, "Don't be a fool, take the position and ask no questions. I know Thomson intimately. He is a queer fish, but ... you can help him and ... he will not be ungrateful." Soon after Haupt accepted the offer, Thomson visited his office, spotted the model of a bridge there, and said, "Some fellow has been trying to make a bridge and he don't know anything about it. He has got his braces in the *wrong* way." Haupt, whose passion was designing bridges, rose to the bait, said the braces were "in the *right* way," and gave his boss a short lecture. Two days later he learned from those under him, not Thomson, that he had been put in charge of planning and building all bridges along the line. Haupt proved to be a superb number-two man, loyal, tireless, and imaginative, but often his thoughts on policy—the policy of the Pennsylvania, he once said, should be "to get all this business that we can conveniently accommodate, and charge upon it as much as it will bear"—had to be translated into more precise, practical terms. How much did it cost to haul twenty bales of cotton in a train that also carried potatoes, wheat, iron, and granite? Slowly, through a steady exchange of ideas, facts, and figures, the roads worked out a cost per ton-mile that covered literally thousands of items. But into every rate variables had to be worked to fit each road's peculiar character. The Pennsylvania for instance, was a mountain road on which the cost per ton-mile was higher than on the Erie or New York Central, both close to water-level routes.

These new founding fathers faced stupendous problems, for the railroad resembled nothing that had previously existed in America. Management of a textile mill, before this time the largest industrial unit, offered no guidance. "Even the biggest of these could be explored, nook and cranny, in an afternoon's leisurely stroll," Harold Livesay remarks. Getting to know even a small railroad, "say thirty or forty miles of route, with such requisite ancillaries as stations, warehouses, repair shops, offices, and roundhouses, took days to tour on foot. Moreover, unlike the textile mill with its five or six working floors lined with machines, only a fraction of the railroad's operations could be seen at one time by a given observer." In the 1850s the largest textile mill employed eight hundred workers at a time when the Pennsylvania had about four thousand people on the payroll. How do you discipline thousands of workers scattered over hundreds of miles, when each can bring an entire operation to a standstill with a large error in judgment? Benjamin Latrobe of the B & O and Daniel McCallum of the Erie published treatises on the subject that were read attentively throughout the industry. The textile owner usually sold his output to a commission merchant who rendered accounts once a month. "Collecting railroad revenues presented crucial and unprecedented problems of management," Livesay goes on. "In sharp contrast, the railroads collected their million-dollar revenues in hundreds of thousands of individual transactions, the great majority of them in cash. It was quite literally a nickel and dime business, most of this cash going first into the hands of conductors and station agents. . . . To ensure that this river of coin flowed into the company's treasury, not into the employees' pockets, required a system of numbered tickets and freight waybills, unique conductors' punches (no two alike and all registered), and station accounts, all supervised by a central office."

The nuts-and-bolts side of railroad technology came with startling speed, largely because of the generous exchange of information among the roads. Except for a standard gauge, "Uniform methods of construction, grading, tunneling, and bridging were developed," Chandler reports. "The iron T rail came into common use. By the late 1840s the locomotive had its cams, sandbox, driver wheels, swivel or bogie truck, and equalizing beams. Passenger coaches had become 'long cars' carrying sixty passengers on reversible seats. Box cars, cattle cars, and other freight cars were smaller but otherwise little different from those used on American roads a century later." Changes in the administration of these burgeoning giants calls for some attention, even though it does not always touch directly upon a history of nuts and bolts.

Haupt and Thomson together created a road on which a line that would soon reach to Chicago was broken down into divisions headed by a superintendent, with a roadmaster in charge of maintenance of the track, a senior mechanic in charge of roundhouses and shops, an assistant trainmaster in charge of yards and the movement of trains. "Thomson's major achievement was to clarify relations between the functional offices of the division and those of central office,'' according to Chandler. He created a line-and-staff organization in which men on the line—those responsible for day-to-day operations—had to report to those on the staff responsible for policy. ''By the coming of the Civil War the modern American business enterprise had appeared among American railroads,'' Chandler remarks. Thomson must have been a dull man to meet face to face, but he left behind a railroad thought to be indestructible until the mid-twentieth century.

Thomson's achievements, among them leading the Pennsylvania Railroad through the Civil War with no public word about an affection for the South, despite a long residence there, can be summarized from Chandlers's study, even though he speaks of the railroad industry generally:

> Improved organization and statistical accounting procedures [many of them inaugurated by Thomson] permitted a more intensive use of available equipment and more speedy delivery of goods by providing a more effective continuous control over all the operations of the road. These innovations also made possible the fuller exploitation of a steadily improving technology which included larger and heavier engines, larger cars, heavier rails, more effective signals, automatic couplers, air brakes, and the like. These improvements permitted the roads to carry a much heavier volume of traffic at higher speeds.

37

America Becomes Aware of Itself

With the railroad, American technology came out of the closet. A few exceptions aside—the grumbling steam pumper of the Philadelphia Waterworks, mass-produced wooden clocks, followed by the omnipresent steamboat, and, on the eve of the Civil War, the growing penumbra of telegraph wires crisscrossing city streets—the nation had little direct, firsthand experience of what technology and industry had been up to since 1800. Machines had long lain hidden in the garden behind the sedate walls of mills and shops, with only the twirling waterwheel to hint at what lay within. The exploitation of anthracite coal and the cheap fuel it gave to stationary steam engines moved the machine from the country to the city; but it still lay hidden behind walls, this time those of the factory. It took the locomotive to tell people what the country's mechanicians and entrepreneurs had been doing for the past half-century.

The plain people loved the railroad passionately—"much as a lover," a French visitor (naturally) said, "loves his mistress." In a similar vein another Frenchman saw the locomotive as the personification of America: "The one seems to hear and understand the other—to have been made for each other—to be indispensable to the other." In small towns the railroad depot came to complement the pleasures offered by the general store and local tavern, a third center of their daily lives. The stationmaster gave from the telegraph the latest news and the disembarking passengers, if there were any, dispensed "vibrations emanating from New York or Chicago." Then there was the pleasure of watching a train making up and starting on its route, an experience that changed little over time. Faulkner's description of such an event in 1942 in a southern village fits what occurred regularly a century earlier:

> Then the little locomotive shrieked and began to move: a rapid churning of exhaust, a lethargic deliberate clashing of slack couplings travelling backward along the train, the exhaust changing to the deep slow clapping bites of power as the caboose too began to move and from the cupola he watched the train's head complete the first and only curve in the entire line's length and vanish into the wilderness, dragging its length of train behind it so that it resembled a small dingy harmless snake vanishing into weeds, drawing him with it too until it ran once more at its maximum clattering speed between the twin walls of unaxed wilderness as of old. It had been harmless once. . . .

Those last words—"it had been harmless once"—underscore the foreboding a number of literary giants of the previous century had felt about the railroad, though not all of them. John Stuart Mill, as paraphrased by Marx, saw the locomotive as "a perfect symbol because its meaning need not be attached to it by a poet; it is inherent in its physical attributes. To see a powerful, efficient machine in the landscape is to know the superiority of the present to the past." Emerson, at least in 1836 in his book *Nature,* felt the same way. He saw man with the railroad changing the face of the Earth. "To diminish friction, he paves the road with iron bars, and, mounting coach with a ship-load of men, animals, and merchandise behind him, he darts through the country, from town to town, like an eagle or a swallow through the air." The whistle of the locomotive comes to him as "music," as "the voice of the civility of the Nineteenth Century saying, 'Here I am.' "

Hawthorne and Thoreau beg to disagree. Long before Emerson began to sense that the locomotive and all it stood for was a "machine that unmans the user," Hawthorne only heard the locomotive shattering a "slumberous peace" in Sleepy Hollow, its whistle not music but a "long shriek, harsh, above all harshness, for the space of a mile cannot mollify it into harmony." Thoreau had a more complicated reaction. He knew the railroad moved men and goods around a constantly expanding country. But he hated the "rattle of railroad cars" that shattered the peace of Walden Pond, and to him the locomotive whistle sounded "like the scream of a hawk sailing over some farmer's yard, informing me that many restless city merchants are arriving within the circle of the town." The locomotive to him snorted "like thunder, shaking the earth with his feet," the iron horse was "devilish," with an "ear-rending neigh . . . heard throughout the town." Ultimately, he condemned the railroad, symbolic for him of a new world he wanted little to do with, in an imperishable phrase— "men have become tools of their tools."

But spokesmen for the plain people and their affection for the railroad prevailed over eccentrics like Thoreau, notably Edward Everett, who was to bore the wits out of an audience the day Lincoln delivered his Gettysburg Address, although he was one of the most popular orators of his day. Everett probably had come no closer to a machine shop than Thoreau, yet he found himself able to extol the American mechanician who, among other things, had built the locomotive. "There is an untold, probably an unimagined, amount of human talent, of high mental power, locked up along the wheels and springs of the machinists; a force of intellect of the loftiest character," he said in one speech. The praise continues in another oration: "He kindles the fires of his steam engine, and the river, the lakes, the ocean, are covered with his flying vessels. . . . He stamps his foot, and a hundred thousand men start into being; not, like those which sprang from the fabled dragon's teeth, armed with weapons for destruction, but furnished with every implement for the service and comfort of man."

Thoreau hated to see the days "minced into hours and fretted by the ticking clock," the peace of his pond shattered by the rattling of railroad cars and the sound of a shrieking locomotive whistle, but he calls for honor as being among the first to sense the ominous aspect of the wave of the future he saw rolling in.

X

The Quality of Life

38

The Home

The amenities that the moderately well-to-do Sidney George Fisher acquired for his home through the nineteenth century—indoor plumbing, central heating, gas lighting, a bathtub, an icebox, a sewing machine—did not filter down to the workingman's family. "The age of invention and mass production," Lewis Mumford has said, "scarcely touched the worker's house or its utilities" until the end of the century. Meanwhile, the workload of the average housewife increased. "Housework," Ruth Schwartz Cowan has observed, was, from the first settlements on, "socially defined as 'women's work,' but in fact the burden was a shared one. The relations between the sexes was reciprocal: women assisted men in the fields and men assisted women in the house." Men drew the water from the creek or well, killed and plucked the chicken that the wife cooked, ground the grain for her bread dough, chopped and brought in wood for the fireplace. Many shared tasks—weaving, milking, paring apples, shucking corn—were "sexually neutral." But as the century wore on and the man who had been in and out or around the house all day left at dawn for the shop or factory and returned after dark, housework became "truly 'women's work'—and not an obligation shared by both sexes." Frances Trollope viewed the workingman's wife as a slave imprisoned in a life "of hardship, privation, and labour" who has lost "every trace of youth and beauty" by the age of thirty.

The one technological innovation most workingmen could offer.

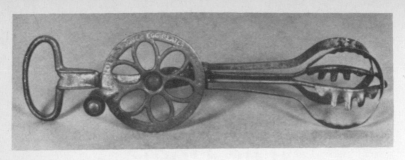

The Dover eggbeater.

their wives was the cast-iron cooking stove, the price of which by the mid-1830s had dropped as low as five dollars. The stove, much as it pleased her, added a new chore to the housewife's schedule. "Get a range as plain as possible," a home economist warned. "Much nickeling, curlicues, scrolls, and gingerbread work generally are a pest to keep clean and often, from an artistic view atrocities. A plain surface is easier to black and clean." With the stove family menus began to expand from the traditional one-pot meals—stews, thick soups, and pottages—into something more varied. While the offerings differed from region to region, within each region a monotonous diet of overcooked meats and vegetables continued to prevail. Local markets did nothing to ease the housewife's burdens. "Shoppers return from the market with live chickens that had to be killed, or dead ones that had to be plucked," Susan Strasser reports. "Even purchased fish had scales; even purchased hams had to be soaked or blanched. Roasting and grinding green coffee, grinding and sifting whole spices, cutting and pounding lump or loaf sugar, sifting heavy flour that might be full of impurities, soaking oatmeal overnight, shelling nuts, grinding cocoa shells, seeding raisins, making and nurturing yeast, drying herbs: tasks like these accompanied nearly every ingredient of every recipe, whether it came from the garden or the market."

In Strasser's apt phrase, "the house shrank during the winter" when most families retreated to the kitchen; then came the annual "household earthquake"—spring cleaning. As one lament runs,

The melancholy days have come
The saddest of the year

When from domestic scenes a man
Will quickly disappear;
For lo! around his humble home
Housecleaning waxeth rife,
And brooms and mops and kindred things
Absorb his wedded wife.

Except for the fortunate home that had a pump handy in the kitchen, every drop of water used—be it for spring cleaning, bathing, cooking, washing dishes or clothes—had to be hauled from a creek or well, then hauled again after using to be dumped outside. On Blue Monday, wash day, "one wash, one boiling, and one rinse used about fifty gallons of water—or four hundred pounds." The weightlifting continued on Tuesday, ironing day. Irons weighed up to ten pounds and had to be constantly reheated on the stove lid. "Even the most pared-down versions of the laundry routine," Strasser says, "demanded enormous amounts of hard, hot, heavy work—hoisting the irons, hauling heavy tubs full of hot water and wet clothes, rubbing, wringing, and bending down for clothes to hang on the line." An extravagance every housewife indulged in when the chance came was to hire a laundress to take charge of the hated routine. The occupation of laundress is one of America's oldest professions.

A family's day began, either before or after breakfast, with someone emptying the chamberpots beneath every bed. Most wives called it "the most disagreeable item in domestic labor." Indoor toilets remained a privilege of the rich into the 1850s, and the privy prevailed in most families until long after the Civil War. In some houses it could be found in a closet off the kitchen—"keeping the window open and the door shut, will prevent any disagreeable effects," one home economist advised—but for most the privy stood like a sentinel in the backyard. Getting there through winter snows called for willpower, and for many, women especially, the urge to delay "their visits to the privy until compelled by unbearable physical discomfort" led them "to become so constipated that days and sometimes weeks will pass between stools."

Every new generation of wives has passed down since the nineteenth century a lament few reasonable husbands have dared to

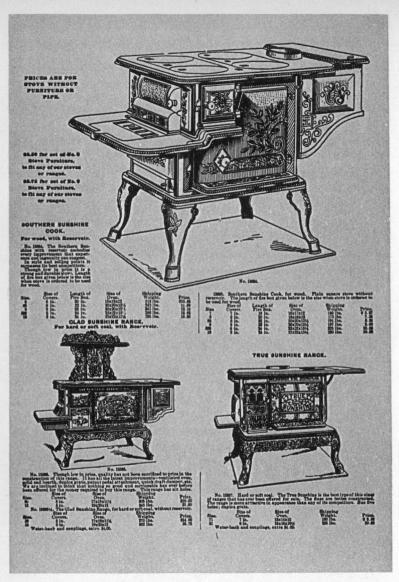

Three "Sunshine" stoves offered by Sears in their 1897 catalog.

dispute—"a woman's work is never done." By the eve of the Civil War wives, rich or poor, had made houses into homes that were all theirs. The home "stood for their values as it stood on the strength of its joists and beams," William Gaddis has remarked. "Cleanliness, order, efficiency and ease—in short, comfort and security—rule this roost." In homes in which affection between the spouses reigned, much of the work was done to please the man who had spent ten to twelve hours away from the home for the welfare of those he found there when he returned, at last, usually long after the sun had sunk. Husbands entered the home after their long day and they "were asked to scrape their boots, smoke in the back, mind their manners, consent to the taming of space and contribute to displays of pots, lamps, lace, and to the collecting of bibelots." They were. They still are.

39

The City

The rise of the factory and railroad, Mumford has said, "transformed the industrial towns into dark hives, busily puffing, clanking, screeching, smoking for twelve and fourteen hours a day, sometimes going around the clock." By the mid-nineteenth century industrialism "had produced the most degraded urban environment the world had yet seen." The description holds for much of Great Britain, but with one major exception it does not hold for America down to the Civil War. The exception was Pittsburgh, endowed with an easy access to soft coal, and which Chevalier in 1838 found "the dirtiest town in the United States." A decade later an English visitor found there every "fresh green leaf and the delicate flower being begrimed, ere they have fully unfolded themselves, by the smoke and soot with which the whole atmosphere is impregnated."

But Pittsburgh was not America. Eastern cities—notably Boston, New York, and Philadelphia—still resembled their colonial origins. As in the past, the stench from decaying horse manure in the streets, from privies in tenement basements, filled the air. Gene Schermerhorn, a youngster in New York in the 1850s, could still rope roving pigs in the street and fish in ponds scattered about Kip's Bay. Hints of the future lay on the horizon. Enormous tanks, for instance, housing illuminating gas "blotted the landscape," Mumford reminds us. "The leakage of escape gas scented the so-called gas-house district, and these districts frequently became among the most degraded sections of the city."

But young Schermerhorn did not recall these eyesores when he reminisced in the 1880s. Uppermost in his mind were days spent flying kites, playing marbles, and "something we called Base Ball were the great games." An easy afternoon buggy ride brought him to villages called Yorkville and Harlem, and to a rural area that had changed little since the Dutch had arrived centuries ago.

Sam Warner has reminded us that the railroad before the Civil War did little to change "the old pattern of the city." It linked towns and cities and brought into urban centers raw goods that left as finished products, but otherwise it had little effect on the daily lives of those who lived in those centers. But the horsecar, an offspring of the railroad, did. It made walking cities, where a man had to live close to his work, into commuting cities. We return to Philadelphia and the man we began with—Sidney George Fisher. He writes in his diary for 1 February 1859 :

> These passenger cars, as they are called, but which are street railroads with horsepower, and which have suddenly sprung into extensive use, are a great convenience. Tho little more than a year old, they have already displaced the heavy, jolting, slow and uncomfortable omnibus and are destined soon to banish it and hacks also entirely. They are roomy, their motion smooth and easy, they are clean, well cushioned and handsome, low to the ground so that it is convenient to get in and out and are driven at a rapid pace. They offer great facilities in traversing the city, now grown so large that the distances are very considerable from place to place.

Fisher always tied technological innovations to the ways they affected him—his comfort, pleasure, or income. The horsecar made it easier to get to his suburban home, and he knew it would in time produce huge advances in real-estate values there. "Every year, almost, now brings with it some new victory of man over nature, of mind over matter, changing the course of civilization and increasing the accommodations of life," he said in 1858 after seeing a machine that threshed grain ; he also said that such machines were "necessary to counteract high wages, which in this country keep down the profits of agriculture." He visited the "hoisting machine" at Philadelphia's Continental Hotel, which took him up to the sixth floor. He departed saying that the hoist "saved a great deal of runnning up and down stairs," but was otherwise unaware that the hotel, a "palace" of ostentation that he denigrated, was enticing him into a new age. "The American hotel," Daniel Boorstin has said, "pioneered the incorporation of mechanical equipment, comforts, and gadgets into architectural designs." They were the first to install indoor plumbing, gaslight (later electric lights), central heating, and the elevator.

40

The Country

Life on a farm changes about as fast as a glacier moves, regardless of the country in which the farmer lives. Alex Shoumatoff recounts a bountiful harvest on a Russian estate in 1906. The Avinoff family had some 5,500 acres of rye, oats, and wheat ready for harvest in August, the busiest month of the year in Russian farmland. "With every horse, wagon, and river in use, it was a bad time to travel. Everyone, even mothers with newborn babes at their breasts, took to the fields, and flashing sickles, accompanied by thrilling songs, hacked down the wheat and corn." Sickles! In 1906, two-thirds of a century after McCormick put his reaper on the market! The harvest over, "the Avinoffs invited forty people to help them celebrate their good fortune," Shoumatoff goes on. "With some of the profits from the harvest, Alexandra Nicolaevna had a set of jewelry made up by Fabergé. The necklace, bracelets, rings, earrings, and brooch were made of golden spikes of wheat, with tiny diamonds as kernels. Mopsy wore the jewels when she was presented at court the following year." No wonder, some would say, the revolution came only a few years later, a remark that overlooks why the Avinoffs refused to buy a reaper for next year's harvest. They had obligations toward the hundreds of peasants on their estate, all of whose families had been serfs not too long ago: the harvest offered them a communal gala and also brought some hard cash into families who seldom saw much of it the rest of the year.

Historians of American technology assume correctly on reading such a tale that it could never happen here. Few American farmers had the time, inclination, or cash to cultivate Fabergé, and none had to worry about the welfare of a mass peasantry; yet most were no more open to technological innovations than Russian landlords. As late as 1850, Cochran notes, ''probably less than a quarter of the farmers of the Northeast used modern methods and equipment.'' Plows with replaceable parts, seed drills, wheel harrows, and mechanical threshers were all available by then, but they seldom as a group turned up on the average farm. The reaper was popular mainly on the flat fields of the Midwest. The hay rake, cheap, made mostly of wood, and simple enough in design for any adept farmer to construct in his work shed, ''was perhaps the major labor-saving device of the age, for it allowed a man, a boy, and a horse to do the work of six men and one or two oxen.''

Agricultural societies flourished in every state and most counties, but gentleman farmers like Sidney George Fisher, who seldom sensed the needs of the typical farmer, ran them and set their tone. In the summer of 1859 Fisher watched the test run of a steam plow promoted by a local society. ''The machine is about the size of a locomotive engine,'' he reported. It pulled behind it eight plows that cut furrows from four inches to a foot deep. It cost $3,500. Fisher admired the work it did, the time and energy it saved, and concluded ''no doubt it will be extensively used in the South and West on large farms and plantations.'' That did not happen until long after the Civil War.

Most cash-poor farmers down to the eve of the Civil War could not afford even manufactured hand tools, let alone steam tractors. It is hard to believe a gentleman Kouwenhoven quotes, who said that ''long before'' 1814, when steel implements were rare, Americans ''left off the use of common iron spades and hoes'' for ones with a blade ''so hard that no stone could injure its edge, and so thin that the spade was driven by hand instead of by foot up to the hub, polished as a razor,'' and that with such a spade he could dig more in one day than two men with iron spades, ''and dance every evening.'' The shadowy history of farm tools reminds us of Fernand Braudel's view that the advance of technology ''is not a linear process.'' Siegfried Giedion at one point remarks that down to 1850, ''as in colonial times, the settlers used wooden implements and utensils, wooden plows, wooden harrows with hickory sticks, mostly of their own fashioning''; then a few pages later he observes that about the same time entrepreneurs, presumably for a profitable market, were producing more than sixty different cast-iron plows, each "shaped for specific purposes, including 'root-breakers, prairie meadow, stubble, selfsharpener, corn, cotton, rice, sugar cane, as well as plows for subsoil and hillside land.'" The single factory-made tool every farmer owned was the ax, which by mid-century had achieved its modern form. An 1848 handbook for emigrants praised America for having "perfected [it] to a high degree. Its curved cutting edge, its heavier head, counterbalanced by the handle, gives the axe greater power in its swing, facilitates its penetration, reduces the expenditure of human energy, speeds up the work.... The handle of the axe is curved, whereby it is more easily guided and more forcibly swung." Another contemporary added that the thin blade "enables it to be more easily drawn out after the blow is given, and the body of the axe, being much firmer, is not liable to twist in working."

In 1858 Emerson gave a talk entitled "The Man with the Hoe." In it he said: "He is permanent, clings to his land as the rocks do. In the town where I live, farms remain in the same families for seven and eight generations; and most of the first settlers, should they reappear on the farms today, would find their own blood and names still in possession." This noble Jeffersonian yeoman was being transformed even as Emerson talked. Countless New England farmers were selling their holdings and moving to the Midwest, where that year McCormick was to sell 4,095 reapers. The American farmer was beginning to "narrow himself to specific products," as a European visitor had remarked four years earlier: "The mass production of apples, peaches, corn, tomatoes, cows, pigs, eggs, or poultry by the American farmer is out of all comparison with the petty European scale." But the great transformation was only beginning before the Civil War, spurred not by the farmer but by entrepreneurs who used technology to create totally new markets for him: food sealed in tin cans, mechanically made bread, refrigerated meat. The farmer's life remained much as it had always been. a dawn-to-dark existence of exhausting manual labor, but inwardly he was becoming something new—an entrepreneur tied to a world technology had imposed upon him.

XI

Crystal Palaces

41

London, 1851

In 1851, prodded by Queen Victoria's consort Prince Albert, Britain invited the world to attend an elegant party in London. She wanted to celebrate the generation of peace that had prevailed since the defeat of Napoleon and also to show off her industrial might. The party would be housed in one of the architectural wonders of the age, a stupendous structure of iron and glass that the press dubbed the Crystal Palace, the perfect sobriquet to transform a "dull industrial exhibition into a cockney fairy tale," as Christopher Hobhouse has put it. A palace was to be built for the plain people, "full of rich stuffs and jewels and lordly ornaments, a paradise of crystal. Incredible as it might seem, Prince Albert's scheme was going to be fun." Seventy-seven countries, among them the United States, accepted the invitation to what was to be the world's first international fair. America, as the new boy on the street of nations, saw a chance to outshine her old enemy, dissolute England. "Can it be possible," a newspaper asked, "that overwhelmed as she is with ... a nest of non-producers in the shape of aristocrats, eating away its vitality, with corruption pervading every fibre and muscle of the body politic, can compete with a young, vigorous, athletic, powerful republic like the United States? We should think not."

The federal government, all but immobilized by the growing division between North and South, refused any financial aid to those eager to exhibit, though it eventually donated a warship to carry their displays across the ocean; an American banker anted up $15,000 to move them from dockside into the palace. Optimists had asked for twice the space needed. Over this desolate "prairie ground," as the British press called the spot, stretched an enormous cardboard eagle, of which *Punch* said, "No eagle, asking of itself where it should dine, and hovering in space without a visible mouthful, could represent the grandeur of contemplative solitude better." A reporter on opening day could not help "being struck by the glaring contrast between the large pretension and little performance, as exemplified in the dreary and empty aspect of the large space claimed by and allotted to America." More than half the exhibits had yet to arrive, and the "few wine-glasses, a square or two of soap, and a pair of salt-cellars" on display did not impress him. Slowly more exhibits trickled in—false teeth, rubber boots, a table of revolvers, another of rifles, McCormick's reaper, tobacco plants, ginned cotton, ice-making machines, telegraph equipment, railroad switches, nautical instruments, artificial eyes and legs, corn-husk mattresses, cast-iron stoves, printing presses, planers, meat biscuits, and so forth—but they lacked the grandeur of the tapestries, sculptures, and other luxurious items found in the neighboring displays of Russia, Italy, and France. Queen Victoria on her tour of the American section said it was "certainly not very interesting," and the South rejoiced in her reaction and those of other visitors. "We too often talk as if we were the only civilized nation on earth," said one southern paper. "It is time that the conceit was taken out of us," it added, meaning by "us" the North. "We of the South, therefore, have no cause for disappointment and may fairly leave that to be monopolized by the elegant manufactures of the North and East."

Despite the South's glee, whoever wrote the catalog notes on the American exhibit made a point about its uniqueness that gradually sank into every visitor to the Crystal Palace:

> The absence in the United States of those vast accumulations of wealth which favour the expenditure of large sums on articles of mere luxury, and the general distribution of the means of procuring the more substantial conveniences of life, impart to the productions of American industry a character distinct from that of many other countries. The expenditure of months or years of labour upon a single article, not to increase its intrinsic

value, but solely to augment its cost or its estimation as an object of *virtu*, is not common in the United States. On the contrary, both manual and mechanical labour are applied with direct reference to increasing the number or the quantity of articles suited to the wants of a whole people and adapted to promote the enjoyment of that moderate competency which prevails among them.

Some of the best of America's song-and-dance men dramatized the catalog's comment on the uniqueness of American technology. The omnipresent Colonel Colt, who urged visitors in the sedate palace to fire his six-shooter, convinced the London *Times* that his "revolvers even threaten to revolutionize military tactics as completely as the original discovery of gunpowder." Others saw his weapon as ideal for subduing "the irregular, impetuous rush of such warlike tribes as the Kaffirs, the Affghans, the American Indians, and the New Zealanders." The impecunious Charles Goodyear (he had no sense of the value of money and probably spent more time in debtor's prison than any well-known American of his day) spread his display of goods made from vulcanized rubber—boats and boots, of course, but also draperies, buttons, inkwells, flooring, to name only a few of the accouterments of the rooms—through a suite that cost $30,000 of most probably somebody else's money to assemble. Gail Borden, soon to market condensed milk in tin cans, offered the public his unappetizing meat biscuit, which he had spent six years creating. They were ideal for men at sea, for travelers "on long journeys" through destitute country, for hospitals, and for all families, "especially in warm weather." Judges of the exhibition said "a more simple, economical, and efficient form of concentrating food has never before been brought before the public," and awarded his innocuous contribution its most coveted prize.

But public acclaim for the American displays came only in late July, when McCormick's reaper went through trials on a farm outside London. The press had ridiculed the beast they had seen on display in the palace as a cross between a flimsy chariot, a treadmill, and something resembling "a flying machine." When it performed admirably on a damp day in a field of wet grain, the English showed they could be as gracious as they had been acerbic. "Gentlemen," said the farmer on whose land the trials had taken place, "let us give the Americans three hearty English cheers." The London *Times* said, "The reaping machine from the United States is the most valuable contribution from abroad to the stock of our previous knowledge that we have yet discovered." Overnight the public attitude toward the American hall changed. "The 'Prairie Ground' is filled with inquirers," an American reported, "and some gentlemen have found out that there are some people who know what they are doing in some other parts of the Globe, as well as this little island."

Meanwhile, the supreme song-and-dance man of the American contingent, Alfred C. Hobbs—a name that rightly belongs to a character in a music comedy—was flabbergasting the British public with a show of his own. Hobbs was a salesman of locks made by a New York City company. He was also probably the world's greatest lockpicker. In public view and in less than half an hour he picked the locks of a British company that had "for years guarded the vaults of the Bank of England." Next he challenged the inviolability of a lock "believed to be as impregnable as Gibraltar," made by Bramah & Co., which for years had offered the large sum of two hundred guineas to anyone who could pick their "unpickable" lock. Hobbs picked it, though it took him fifty-one hours spread over nearly a month. Once again the British were gracious. The Bank of England ordered a set of locks from Hobbs's company, and the *Times* admitted that "our descendants on the other side of the water are even now and then administering to the mother country a wholesome filial lesson . . . and recently they have been 'rubbing us up' with a severity which perhaps we merited for sneering at their shortcomings in the Exhibition."

Icing on the cake came a few days after Hobbs's triumph, when the yacht *America* beat the British entry *Titania* in a race known today as the "America's Cup." The triumph "has created a positive furor in England," the press reported. The *America* "has beaten everything and borne away the laurels of victory from vessels on whose construction the greatest pains have been bestowed, on whose outfit thousands of pounds have been expended, and in whose success the owners felt necessarily a personal as well as a national pride." *Punch* reacted to the event with a want ad:

"Timber for Sale—A great quantity of Planks, Sticks, Masts and Spars, to be had Cheap.—Inquire at the Royal Yacht Club House, Cowes."

Through all of this hoopla a display by a group of quiet, possibly dour, Vermont rifle-makers from the firm of Robbins & Lawrence caught little attention. Their rifles were among the first produced by private enterprise in America that could be taken apart with a screwdriver and if spare parts were handy put in working order again with a few twists. All its parts were interchangeable. Mechanicians and entrepreneurs who made rifles were impressed, and what they saw helped to convince Parliament to send a commission to find out what America was up to. (The commission arrived as the Crimean War ended an age of peace, and on its recommendation Britain awarded Robbins & Lawrence with an order for 25,000 rifles with interchangeable parts.)

Subtleties of this sort were missed by the British press. *Punch* honored the "whack" of Colonel Colt's revolver, but the rifle with interchangeable parts went unnoticed in the doggerel that gave a gracious tip of the hat to America:

> *Yankee Doodle sent to town*
> *His goods for exhibition;*
> *Everybody ran him down*
> *And laughed at his position;*
> *They thought him all the world behind*
> *A gooney, muff, or noodle.*
> *Laugh on, good people—never mind—*
> *Says quiet Yankee Doodle.*

42

New York City—1853

A politician faced with a crisis is said to have told his staff he did not believe in unmitigated disasters. "Now," he added, "go out and mitigate this one." The New York Crystal Palace Exhibition of 1853 was a mitigated disaster. Even P. T. Barnum, called in at a late date to keep the fair alive as a permanent show, confessed, "I was an ass for having anything to do with the Crystal Palace." Except for an excellent essay by Charles Hirschfeld, scholars have paid almost no attention to the disaster. Its promoters located it on a site then regarded as being in the boondocks, the corner of Forty-second Street and Fifth Avenue, adjacent to the Croton Reservoir, where Bryant Park is today. The fair opened late, and despite the ballyhoo of two self-appointed publicity agents—Horace Greeley and Walt Whitman —drew disappointing crowds. It did, though, make a large and lasting contribution to other world fairs that would proliferate through the rest of the century—boulevards of sideshows, gambling dens, cockfight arenas, and, as Hirschfeld reports, other "haunts of dissipation. Here visitors who were looking for something more entertaining than water-pumps or plows could have their fill of 'double-headed calves and harlequin performance,' five-legged cows, dancing bears, mermaids, dwarfs, giants, rattle-

The New York Crystal Palace. *(New York Public Library, Astor, Lenox, and Tilden Foundations)*

256

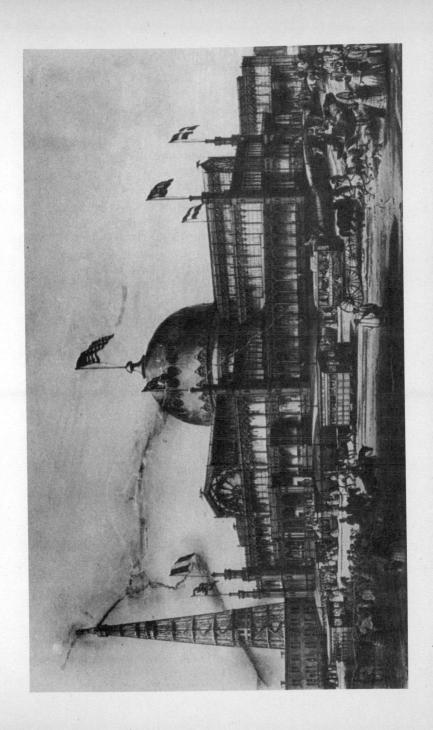

snakes, and 'grinning darkies,' the whole array of popular attractions that have crowded midways of international fairs ever since.''

Among these diversions, a notable song-and-dance man—Elisha Graves Otis, reared on a Vermont farm—had installed an attraction as popular as the parachute jump at the 1939 New York fair, an elevator that rose to the top of a tower nearly three hundred feet high and offered a bird's-eye view of the city. Otis reassured tremulous patrons with a brief drama. He had, in the words of Boorstin, ''set up ratchets along each side of the shaft and attached teeth to the sides of the cage. These teeth were held

The first Otis elevator was displayed at the Crystal Palace exposition in 1853 by Elisha Otis, who after being hoisted to the ceiling had the rope cut and descended to safety. *(New York Public Library)*

clear of the ratchets by the rope which held up the cage, but when the rope ceased to be in tension, the teeth were released against the sides of the shaft and gripped the cage safely in place." As the elevator slowly rose, Otis "melodramatically cut the supporting ropes and displayed himself in the cage safely held in place."

The promoters, who sold stock in the enterprise, were accused of using the exhibition to make money for themselves. They turned down an innovative design for the main hall for one that mimicked London's Crystal Palace and which Giedion has dismissed as "mediocre" but pleased Whitman immensely. He called it "Earth's modern wonder,"

> *Mightier than Egypt's tombs,*
> *Fairer than Grecia's, Roma's temples,*
> *Prouder than Milan's statued, spired cathedral.*

Costs soared from the allotted $200,000 to more than $600,000 by opening day. A British commission that came in June found that day had been postponed, and another month passed before it finally arrived. Even then less than one-third of the displays were in place. The press told the public to stay away, and for the most part it did until early September, when Horace Greeley reassured the nation that all was in order at last. By then even the roof, which had leaked badly, had been repaired.

Whitman's "Song of the Exposition" gave a brief but precise picture of what visitors would see in the palace. "The cotton shall be pick'd almost in the very field, / Shall be dried, clean'd, ginn'd, baled, spun into thread and cloth before you." Crude ore from mines in California and Nevada would be transformed before the visitor's eyes.

> *You shall mark in amazement the Hoe press whirling its*
> *cylinders,*
> *shedding the printed leaves steady and fast,*
> *The photograph, model, watch, pin, nail, shall be created*
> *before you.*

He went on to urge visitors toward "the music house" and the halls that held sculptures and paintings. All the displays sought "to teach the average man the glory of his daily walk and trade," he said.

(This, this and these, America, shall be your *pyramids and*
  obelisks,
Your Alexandrian Pharos, gardens of Babylon,
Your temple at Olympia.)

A reporter for the *Illustrated News* of New York offered a
high-class entry for the worst piece of writing in a summer of
flamboyant statements about the fair. The Crystal Palace, he said,
gave to the world "a sweet sunprint of the glowing Future—
science and art in an amicable wrestle for the smile of beauty; the
loom and the anvil laughing out the jocund sound of profitable
labor; the steam-engine snorting its song of speed; the telegraph
flashing its words of living flame; the subdued ocean bridged with
golden boats. . . . In the poetry of notable events the Crystal Palace may be termed the Iliad of the Nineteenth Century. In this
country its Homer should be the people."

A hope prevailed in the year after *Uncle Tom's Cabin* had been
published that a fair participated in by all thirty-one states would
help to paper over differences between the North and the South.
Much of the machinery built in the North was praised for its lack
of "tinsel, or extra gew-gaw show, such as some exhibitors seem
to think make their articles more attractive." The South mostly
contributed displays that came from the farm, but it did send in
the Southern Belle, a stream engine built by the Winter Iron
Works of Montgomery, Alabama. It ran all the machines in the

Roebling's Niagara River Bridge. This illustration serves as a fitting end
to our portfolio, for it marks the wave of the future in American technology, a wave that culminated some thirty years later in the opening of
John A. Roebling's Brooklyn Bridge. The 821-foot span over the Niagara
River, just above the Falls, carried a railroad on the upper level and a
carriageway and walkway on the lower one. The bridge, in the words of
David Billington, "confounded nearly all engineering judgment of the
age, which held that suspension bridges could never sustain railway
traffic." It was torn down in 1897, forty-two years after it opened, when
maintenance costs—Roebling had been forced to use wood to save expenses rather than the iron or steel he wanted—and heavier train load-
ings made it impractical to keep it in operation. *(New-York Historical
Society)*

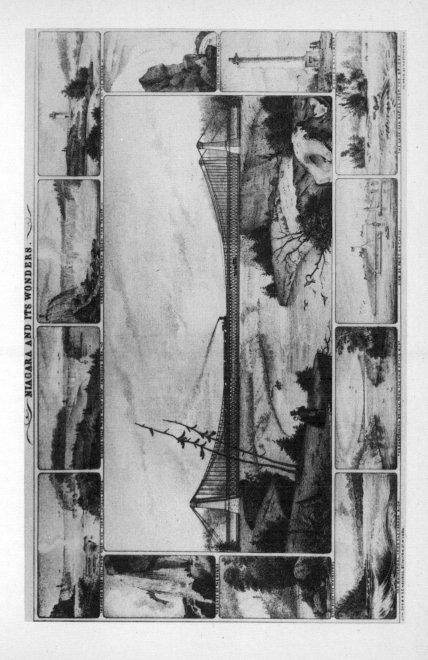

NIAGARA AND ITS WONDERS.

Hall of Machinery. As far as is known, it did its work well, but someone (probably a northerner) had to say, "The Southern Belle, running without any labor, is true to the name *belle*—very showy, and (at present) very useless. No shop would ever dream of making or buying such an engine for use. It would keep one man busy the whole time just to keep it bright and clean."

In a bit over two months the fair packed in more than a million visitors who paid more than $330,000 in entrance fees, far short of the promoters' basic expenses. Let Hirschfeld report the swift and sad end: "After October 1853, as winter came on, the bloom faded quickly. The people stayed away once more. Debts mounted and the stock fell till it was worthless. The directors, in desperation, decided to carry on into the winter, and had stoves installed. There was talk of making the Exhibition permanent in the hope of eventually reclaiming losses." P. T. Barnum was asked to take charge of the dying beast. He soon confessed "the dead could not be raised," and in November the directors declared bankruptcy. And that was the end of the Crystal Palace Exhibition in New York City in 1853. The palace itself burned to the ground four years later.

At the time the exhibition died, America's attitude, still somewhat innocent, still bitterly anti-British, resembled that in the opening line of a song Irving Berlin would write a century later: "Anything you can do I can do better." Some things it could, but Great Britain was still the world's leading workshop. America's achievements, whether based on what it had stolen, borrowed, or invented, reflected, often in bombast, the still flourishing Spirit of '76, and those achievements were remarkable considering the brief span of years it had taken for them to emerge from the shops of its mechanicians and song-and-dance men. Nonetheless, it was to take another quarter-century, until the great Centennial of 1876, for American boasts to be based on fact that in the world of nuts and bolts it truly led all nations.

Notes

Articles and books referred to in these notes are given a full citation when mentioned only once. Otherwise a succinct notation—for example, Hindle, *Emulation*, p. 69, to identify one of several works by Hindle, or Sellers, p. 118, when only a single work by the author has been used—is amplified in the bibliography that follows these notes.

Opening Epigraphs

PAGE

xiii. Søren Kierkegaard quoted in John Mortimer, *Paradise Postponed* (1986), p. 140.

xiii. Robertson Davies, *The Rebel Angels* (1982), p. 169.

Introduction: A Cautious Approach to a Difficult Subject— Some of the Pitfalls That Lie Ahead

1. Fisher, *Philadelphia Perspective*: on Wetherill, p. 74; and indoor plumbing, pp. 240–241; on hoisting machine, p. 403; on sewing machine, p. 369.

2. Ferguson on secrecy: editorial headnote in Sellers, p. 116.

3. "Every tooth": Sellers, p. 168; on Evans's carding machine, ibid.

3. "he made redundant a Boston factory,": in Ferguson, "Americanness of American Technology," p. 8; Allen's description of Whittemore's machine quoted in Rosenberg, "America's Rise to Woodworking Leadership," pp. 200–201; Habakkuk, pp. 93n, 121; Burlingame, *Iron Men*, pp. 140–141.

4. Brandt: see Sellers, pp. 169–170.

4. "One cannot blame the industrialist": Giedion, p. vi.

5. "a great deal that was bad": Chevalier, pp. 106–107.

5. Hogs in Cincinnati: Trollope, p. 89; Martineau, *Western Travels*, vol. 2, p. 45; S. W. Jackman, ed., *Captain Frederick Marryat, A Diary in America, with Remarks on Its Institutions* (1962), pp. 222–223; Giedion, p. 95.

7. "there is no single onward movement": Braudel, p. 334.

7. Cooper on ploughs: *Notions of Americans* (1828), quoted in Sanford, p. 393.

7. Bigelow and technology: Marx, p. 149, 376; Struik, p. 170; Perry Miller, *The Life of the Mind in America from the Revolution to the Civil War* (1965), p. 289; Noble, *America by Design*, p. 5.

8. Definitions of technology: Braudel, p. 430; Layton quoted in Jeremy, p. 4.

8. "Technology is the making of things that did not previously exist.": Billington, *The Tower and the Bridge* (1983), p. 9.

Chapter 1: A Backward Glance

9. "The newly born": Hindle, *Emulation*, p. 2.

12. "to introduce a new tool": Tarule, p. 30.

12. "of good quality and well seasoned": Hummel, p. 44.

16. "were more apt to restrict": Wallace, *Innovation*, p. 152.

16. Peale and Fitch backgrounds: Hindle, essay on Peale, p. 108.

16. "generally worked as an individual": Bedini, "Artisans in Wood," p. 111.

16. "is a mechanic by nature": Chevalier quoted in Hindle, *Emulation*, pp. 133–134.

16. "his own methods": Habakkuk quoted in Fisher, *Workshops*, p. 199.

17. "Those who labor in the earth": Jefferson quoted in Kasson, p. 16.

17. "sponge upon the earth": Beverley, p. 171.

17. Crèvecoeur quotes: pp. 143, 71.

18. Klinkenborg quotes: pp. 7, 33–34.

18. "there was an almost complete dearth of skilled artisans": Burlingame, *Machines That Built America*, p. 13.

18. Bridenbaugh quotes: *Colonial Craftsman*, pp. 31, 46.

Chapter 2: The Spirit of '76

20. "When fifty or sixty men": quoted in Hawke, p. 6.

20. "powder manufactured by Mr. O. Eve's mill": Hawke, p. 31.

20. "a society for the improvement of agriculture": quoted in Kasson, p. 11.

20. On Francis Lewis see David Freeman Hawke, *Honorable Treason: The Declaration of Independence and the Men Who Signed It* (1976), p. 100.

20. "As Congress is now a-promoting": Gilpin to Franklin, 29 December 1775, *Franklin Papers*, vol. 22, p. 321.

22. On Nathan Sellers: Sellers, p. 89.

22. On Leddeus Dod: Bedini, *Thinkers and Tinkers*, p. 241.

22. On Revere: Forbes, pp. 317, 302–303; Oliver, p. 97.

22. "the gunsmith's business": Franklin to Deane, *Franklin Papers*, vol. 22, p. 184.

23. "Agreed that it be recommended": *Franklin Papers*, vol. 22, p. 183.

23. "demands made by the war": Oliver, p. 89.

23. "undoubtedly one of the greatest": Flexner, p. 220.

23. Experiments of Rittenhouse, et al.: Hindle, *Pursuit of Science*, pp. 125–130; Kasson, pp. 12–14.

23. "We are in great want of good engineers": Franklin to Charles-Guillaume-Frédéric Dumas, 9 December 1775, *Franklin Papers*, vol. 22, p. 290.

24. "We want coals": The Committee of Correspondence to Silas Deane, 1 October 1776, *Franklin Papers*, vol. 22, p. 643.

24. "Great scenes inspire great ideas": quoted in Struik, p. 103.

26. "the greatest technological advance": Cochran, p. 58.

26. "the unity of no other nation": Kouwenhoven, p. 42.

26. "Among the causes which gave the impetus": Quincy quoted in Struik, p. 93.

26. "Nothing so clearly documents": Bernard Bailyn, "The Central Themes of the American Revolution: An Interpretation," in *Essays on the American Revolution*, edited by Stephen G. Kurtz and James H. Hutson (1973), p. 20.

Chapter 3: The Founding Fathers

27. "I worry": John Newhouse, *A Sporty Game* (1982), p. 83.

27. "engineering schools": Billington, p. 277, note 7.

27. "considered only an overseer": Henry Adams, vol. 1, p. 130.

28. "We go forward without facts": quoted in Halberstam, pp. 69, 87.

28. "is in large part intellectual work": Wallace, *Rockdale*, p. 237.

28. "Nathan, I have no doubt": Sellers, p. 34.

28. "The craftsman isn't ever following": Pirsig, p. 148.

28. "My own eyes know": Sturt, p. 24.

29. "chief artist": quoted in Sanford, p. 373.

29. "art": Pirsig, p. 260.

30. "Science?": Sturt, p. 19.

30. Walter Chrysler: quoted in Halberstam, p. 50.

30. "On one of them was half": Sellers, p. 49.

30. "Father was very ready": Sellers, p. 49.

30. "Mr. Ford first sketched out": Lacey, pp. 91, 90.

31. Mechanician defined: *Oxford English Dictionary*.

31. "remains one of the few": Hans Zinsser, *Rats, Lice and History* (1935), p. 9.

32. "to enlarge his knowledge": Sellers, p. 52.

32. "in a feverish state of excitement": Sellers, p. 133.

32. "a skill for making just about": Landes, pp. 204–205.

33. Condorcet: quoted in Boorstin, *Discoverers*, p. 67.

33. "a slow accretion": Braudel, p. 334.

34. Eckfeldt and Saxton: Sellers, pp. 53, 70.

Chapter 4: The Making of a Mechanician

35. "I never can forget": Sellers, pp. 29–30.

36. "and long discussions came": Sellers, p. 30.

36. "the more move the more learn": Sellers, p. 46.

36. "many good lessons": Sellers, p. 70.

36. "Seeing me peering over the bar": Sellers, pp. 64–66.

37. "It is curious": Sellers, pp. 43–44.

37. "enduring physical facilities": Wallace, *Innovation*, p. 101.

38. Pat Lyon's office: Sellers, p. 4.

38. "With all this scattered work": Sellers, p. 44.

38. "Find out what was wrong": Halberstam, p. 269.

38. "It is this understanding of Quality": Pirsig, p. 257.

38. "If you have a high evaluation": Pirsig, p. 283.

39. "He used to tell us": Sellers, p. 30.

Chapter 5 : Fellow Travelers

40. "The machinist thought with his hands": Wallace, *Rockdale*, p. 112.

41. "It seems to me almost a miracle": quoted in Wallace, *Rockdale*, p. 114.

41. "It was Coxe": Marx, pp. 155–156.

42. "repository of machines": quoted in Hindle, essay on Peale, p. 152.

42. "well executed in point of workmanship": Washington to Gouverneur Morris, 28 November 1788, quoted in Landes, p. 265.

42. "will lead us into virtue": quoted in Kasson, p. 31.

42. "You know I had a wagon": quoted in Hawke, *Transaction*, pp. 46–47.

42. "a walking stick": Martin, p. 86.

43. "a country where there is more": quoted in Martin, p. 68.

43. "how useful the invention": Peterson, p. 59.

43. "This machine": Martin, p. 102.

43. "my household manufactures": quoted in Martin, p. 98.

43. "had no choice": Kasson, p. 24.

44. "large and expensive machinery": Martin, p. 98.

44. "to the compass of a private family": Martin, p. 98.

44. "the machine is a token": Marx, p. 15.

Chapter 6 : The Brain Drain Begins

48. "We are destitute": Burlingame, *Machines That Built America*, pp. 42–43.

49. "are good for nothing": Burlingame, *Machines That Built America*, p. 43.

50. "A country less developed": Hindle and Lubar, pp. 61–62.

50. On the "Old Slater Mill": see Kulik, pp. 300–334.

51. "A new animation": Dwight, vol. 4, p. 348.

52. "he acted as the node": Jeremy, p. 87.

52. "a close copier": Jeremy, p. 83.

52. "may well have been the first": Penn, pp. 56–66. See note 3, pp. 57–58 for a bibliography of the work done on David Wilkinson.

53. "In Wilkinson's shop was trained": Woodbury, p. 92.

54. "Englishmen knows every thing better": quoted in Jeremy, p. 131.

54. "This establishment suffered much": quoted in Jeremy, p. 90.

54. "What can you expect": Leland, p. 26.

54. "possessed obsolete skills": Jeremy, p. 160.

55. "I cannot conceive": Jeremy, p. 130.

Chapter 7: A Practical Visionary

56. "a plain mechanical-minded": Evans to Aaron Ogden, 8 August 1817, quoted in Sanford, p. 374.

56. "if an open glass be filled": quoted in Giedion, p. 599.

56. "The first precise scientific vision": Giedion, p. 599.

57. Family on Evans: quoted in Burlingame, *Machines That Built America*, p. 23.

57. Brissot de Warville on Evans's mill: Brissot de Warville, p. 163.

58. "who first incorporated the three": Giedion, p. 78.

58. "the great quantity of dirt": quoted in Burlingame, *Machines That Built America*, p. 26.

59. "I have walked through mills": quoted in Burlingame, *Machines That Built America*, p. 29.

59. "flour milling was the leading": Cowan, p. 48.

60. "with contempt little short of insult": Sellers, p. 37.

60. "He set the dredge waddling": Flexner, p. 237.

60. Latrobe on Evans's dredge: Sellers, p. 37.

60. "not only navigating our rivers": Sellers, p. 36.

61. "Mr. Evans had much to say": Sellers, p. 38.

61. "to collect and publish": Sellers, p. 38.

61. "If a scholar poses": Condorcet quoted in Boorstin, *Discoverers*, p. 67.

61. "the workshops": Condorcet quoted in Boorstin, *Discoverers*, p. 67.

62. "vomiting its wreath": Sellers, p. 38.

62. "The time will come": Sellers, p. 38.

Chapter 8: A Ludicrous Yoking of Old and New

63. "an unprecedented number": Mease, p. 147.

63. "a copious supply of water": Mease, p. 147.

64. "a steam engine is": Stapleton, p. 39.

64. "proved too much": Sellers, p. 37.

64. "artistically finished drawings": Sellers, p. 38.

65. "a ludicrous yoking": Sanford, p. 39.

65. "over every other spot": quoted in David Hawke, *In the Midst of a Revolution* (1961), p. 40.

65. "Latrobe represented": Sanford, p. 429.

65. "the first water was thrown": Mease, p. 149.

66. "ran a deficit": Peterson, p. 741.

66. "ignorant of his business": Cope, pp. 80, 60, and Roosevelt's cost overrun, p. 43.

67. "became the most celebrated": Stapleton, p. 42.

67. "The lower engine": Mease, p. 151.

67. "The air pump": Mease, p. 151.

67. Water pipes: Mease, pp. 151, 152.

68. "at night": Sellers, p. 15.

Chapter 9: A Gadget for the Plain People

69. "There, too, the craftsman awoke": Landes, p. 25.

69. "The first thing one has": Landes, p. 25.

69. "Wherever we have been in Kentucky": quoted in Burlingame, *Machines That Built America*, p. 94.

71. "nothing ever has been effective": Braudel, p. 311.

71. "It was this factory": Landes, p. 311.

72. "was both a stylistic and": Hindle and Lubar, p. 222.

73. Peddler-farmer dialogue: Burlingame, *Machines That Built America*, pp. 94–95.

74. "The clock is not merely": Lewis Mumford quoted as epigraph in Landes, p. xix.

Chapter 10: Steamboats in the East

75. "separated the colonial": Adams, vol. 4, p. 135.

75. "not only the first steamboat": Flexner, p. 115.

76. "a thing of beauty": Sturt, p. 73.

76. "a bewildering mixture": Newhouse, p. 74.

76. "Approaching a machine-age problem": Flexner, p. 112.

77. "Every artist who invents": Fulton to Aaron Ogden, 1813, quoted in Sanford, pp. 369, 366.

77. "that every aspect of submarine warfare": Flexner, p. 227.

78. "the most original": Jefferson quoted in Flexner, p. 149.

78. "Pioneering don't pay": Carnegie quoted in Morison, *From Know-How to Nowhere*, p. 115.

78. "the first mechanical genius": Voight quoted in Hindle, *Emulation*, p. 36; also Flexner, p. 179.

79. "Clockmakers were the finest mechanics available": Hindle, *Emulation*, p. 37.

79. Launching of Fitch boat: Brissot de Warville, p. 195.

79. "might have been a fashionable picnic": Flexner, p. 282.

80. "with a view to the navigation of the Mississippi": Flexner, p. 280.

80. "into a floating palace": Flexner, p. 289.

80. "Went to Newport on Thursday": Fisher, *Philadelphia Perspective*, p. 198.

Chapter 11: Steamboats on Western Waters

82. "discovered the powerful principles": Evans to Timothy Pitkin, 21 December 1811, Sanford, pp. 372–373.

82. "I began to try to induce people": Sanford, p. 373.

82. "Thousands of spectators": Sanford, p. 372.

83. "noncondensing, direct-acting": Hunter, "Invention of the Western Steamboat," p. 39.

84. "an engine on a raft": Hunter, *Steamboats*, p. 62.

84. "must be so built": Hunter, *Steamboats*, p. 84.

84. "tin, shingles": Hunter, *Steamboats*, p. 63n.

84. "If a steamboat should go to sea": Hunter, *Steamboats*, p. 63n.

84. "The western steamboats": Chevalier, p. 217.

84. "very fine and comfortable": George Wilson Pierson, *Tocqueville in America*, abridged ed. (1959), p. 364.

84. "no two steamboats are alike": David Stephenson, quoted in Kouwenhoven, p. 28.

85. "of a boat about": Chevalier, in letter dated New Orleans, 8 January 1835, p. 218.

85. The history of the steamboat: Hunter, *Steamboats*, pp. 121–122.

85. "notwithstanding the elegance": Chevalier, p. 217.

85. "Talk about *Northern* steamers": Boorstin, *National Experience*, p. 100.

86. "were huge trees": John McPhee, "The Control of Nature: Atchafalaya," *The New Yorker*, 23 February 1987: 55.

86. "The suggestion of the method": Hunter, "Invention of the Western Steamboat," p. 45.

86. "if the returns were all in": Hunter, "Invention of the Western Steamboat," p. 46.

86. "The story of the evolution": Hunter, *Steamboats*, pp. 121–122.

Chapter 12: A Wave of the Future

88. "You tell me I am quoted": Jefferson to Benjamin Austin, 9 January 1816, in *Portable Thomas Jefferson*, edited by Merrill D. Peterson (1977), pp. 447–450.

89. "The first measure": Nathan Appleton, "Introduction of Power Loom" (1858), quoted in Sanford, p. 353.

91. "is unquestionably entitled": Sanford, p. 357.

91. A New Hampshire millowner: quoted in Jeremy, p. 169.

91. "an extemporaneous town": Chevalier, p. 136.

91. "was the informing soul": Appleton quoted in Sanford, p. 357.

92. Dialogue with Shepard: Appleton, quoted in Sanford, p. 354.

92. "necessitated higher operating speeds": Jeremy, p. 190.

92. "invented a machine for warping": quoted in Jeremy, p. 191.

93. "Should I be fortunate enough": Struik, p. 146.

Chapter 13: A Song-and-Dance Man?

95. "The year 1815 was": Smith, "Military Entrepreneurship," pp. 64–65.

97. Song-and-dance man defined: See *Dictonary of American Slang*, compiled by Harold Wentworth and Stuart Berg Flexner (1960).

99. "I am persuaded": quoted in Hounshell, *American System*, p. 30.

99. 'My intention is": Nevins and Mirsky, p. 190.

100. "to substitute correct": Kouwenhoven, p. 34.

100. "Nathan, when I hire a workman": Sellers, p. 34.

100. Tocqueville on Whitney's vision: quoted in Fisher, *Workshops*, p. 34; see also pp. 139, 141–143.

101. "An improvement is made": Jefferson to John Jay, 30 August 1785, quoted in Hounshell, *American System*, p. 26. The best source on Blanc is still the long-forgotten article of W. F. Durfee, "The First Systematic Attempt at Interchangeability in Firearms," *Cassier's Magazine* 5 (April 1904): 469–477. On the link to Jefferson and Whitney, see Lionel T. C. Rolt, *A Short History of Machine Tools* (1965), p. 139–141; and Nevins and Mirsky, chap. 5: "Jefferson and a Practical Demonstration."

101. "Mr. Whitney has invented": Smith, "Whitney," p. 47.

101. "Artist of his Country": Smith, "Whitney," p. 47.

102. "Whitney's reputation as inventor": Smith, "Whitney," pp. 48–49.

102. "Among these pathfinders": Livesay, *American Made*, p. 22.

Chapter 14: The Springfield Armory

104. "Fighting among the workmen": regulations quoted in Hindle and Lubar, p. 233.

104. "some forty or fifty": Uselding, "Technical Progress," p. 76.

104. "There are very few in the armory": Uselding, "Technical Progress," p. 76.

104. "all the masters": Uselding, "Technical Progress," p. 75.

105. "to draw up a system of regulations": Smith, "Military Entrepreneurship," p. 65.

105. "Being told what was wanted": Smith, *Harpers Ferry*, p. 124.

105. "Awkward looking but amazingly efficient": Smith, *Harpers Ferry*, p. 127.

106. "Rarely have the contributions": Smith, *Harpers Ferry*, p. 138.

106. "Blanchard clearly did not": Hounshell, *American System*, p. 38.

108. "had reached a point of sophistication": Smith, *Harpers Ferry*, p. 199.

108. "The master armorer had a set": Smith, *Harpers Ferry*, pp. 199–200.

108. "to make the muskets": Green, p. 172, quoting letter to Lee dated June 1819.

Chapter 15: A Visionary Theorist

109. "a visionary theorist": Hounshell, *American System*, p. 42; also Smith, *Harpers Ferry*, p. 156.

110. "if a thousand guns were taken apart": Hall to Bomford, 18 January 1816, quoted in Smith, *Harpers Ferry*, p. 191.

111. "will cost the government": Smith, *Harpers Ferry*, p. 198.

111. "will answer as well": Hall to Calhoun, 15 May 1822, Smith, *Harpers Ferry*, p. 199.

111. "I have succeeded": Hall to Calhoun, 30 December 1822, Smith, *Harpers Ferry*, p. 199.

111. "more numerous and exact": Hounshell, *American System*, p. 41.

111. "became one of the fundamental": Hounshell, *American System*, p. 41.

112. "practical judges of machinery": Smith, *Harpers Ferry*, p. 202.

112. "By no other process": Smith, *Harpers Ferry*, p. 202.

112. "All told, Hall emerges": Smith, *Harpers Ferry*, p. 249.

113. "added another dimension": Smith, "Military Enterpreneurship," pp. 76–77.

113. "The development of the American System": Hounshell, *American System*, p. 44.

Chapter 16: Why Not the South?

117. "Tho' their country be overrun": Robert Beverley, *The History and Present State of Virginia* (1705; reprinted 1947), p. 295.

117. "The grave was dug": Henry Grady, editor, of *Atlanta Constitution*, quoted in Jacobs, p. 41.

118. Appleton on Calhoun: quoted in Sanford, p. 359.

119. On Findlay: see Davis, pp. 25–28.

119. "the slave states produced": Eaton, p. 246.

119. "seemed least appreciated": Neil Harris, *Humbug: The Art of P. T. Barnum* (1973), p. 84.

120. "broke the opposition": quoted in Bateman and Weiss, p. 22.

120. "The agrarian ideal": Eaton, p. 244.

120. "lived by being complaisant": Eaton, p. 244.

120. "The tremendous improvement": Lindstrom, "Industrialization of the East," pp. 39–40.

120. "has prevented our perfecting the arrangements": quoted in Calhoun, p. 85.

121. On Pratt and Thomson: see Eaton, pp. 241–242.

121. "growth poles" defined: Lindstrom, "Industrialization of the East," p. 34.

121. "Louisville is at this time": quoted in Sinclair, pp. 13–14.

122. "An exclusively agricultural people": quoted in Thomas C. Cochran and William Miller, *The Age of Enterprise: A Social History of Industrial America* (revised edition, 1961), p. 33.

122. "A small shop producing transmission": Jacobs, p. 41.

122. "It was a great old shop": Henry Strauss quoted in Lacey, p. 23.

123. "were cast at Park's foundry": Sellers, p. 107.

Chapter 17: Machines in the Garden

124. "a harmony of rural and industrial interests": Wallace, *Rockdale*, pp. 4–5.

124. Zachariah Allen: quoted in Hunter, "Waterpower," pp. 178–179.

124. "to find 50": Hunter, "Waterpower," p. 179.

125. "Observations": noted in Wallace, *Rockdale*, pp. 114–115.

125. "In the 1820s": J. W. Lozier, "The Forgotten Industry: Small and Medium Sized Cotton Mills South of Boston," *Eleutherian Mills Working Papers* 2 (1979), no. 4: 119–120.

127. "did not necessarily lead": Betsy Bahr, "The Antietam Woolen Manufacturing Company: A Case Study in American Industrial Beginnings," *Eleutherian Mills Working Papers* 4 (1981), no. 4: 36.

127. "a little nausea": Wallace, *Rockdale*, p. 182.

127. "did not find it necessary": William A. Sisson, "From Farm to

Factory: Work Values and Discipline in Two Early Textile Mills,"
Eleutherian Mills Working Papers 4 (1981), no. 4: 20.

128. "was shut down forty-one days": J. W. Lozier, "Rural Textile Mill
Communities and the Transition to Industrialism in America, 1800–
1840," *Eleutherian Mills Working Papers* 4 (1981), no. 4: 87.

128. "truly poor": Billy G. Smith, "Struggles of the 'Lower Sort' in
Late 18th Century Philadelphia," *Eleutherian Mills Working Papers* 2
(1979), no. 1: 17.

128. "The general picture which emerges": Donald R. Adams, Jr.,
"Workers on the Brandywine: The Response to Early Industrialism,"
Eleutherian Mills Working Papers 3 (1980), no. 4: 10–11.

129. Cost of moving to Ohio: Wallace, *Rockdale*, p. 65.

Chapter 18: The Machine Shop

130. "a born promoter": Noble, *Forces of Production,* pp. 96, 195.

130. "with his head close to the shaft": Sellers, p. 107.

131. Milling machine: Hounshell, *American System,* p. 29.

131. Tracer control: Noble, *Forces of Production,* p. 82.

132. "is justly celebrated": quoted in Woodbury, p. 75.

132. "A machine then was clearly a machine": "Talk of the Town"
column, *The New Yorker,* 15 August 1983: 27.

132. "its own unique personality": Pirsig, p. 39.

133. "own exacting standards": Rolt, p. 14.

133. "My vision of the possibilities": Leland to Freeman, 11 February
1927, quoted in Hounshell, *American System,* p. 81.

133. "if one is making a wheel barrow": Leland, p. 45.

Chapter 19: The Southerner in the North

137. "the expense of drenching reapers": Rush "To the Editor of *The
Pennsylvania Journal:* Against Spirituous Liquors," 22 June 1782, in
The Letters of Benjamin Rush, edited by Lyman H. Butterfield, 2 vols.
(1951), p. 271.

138. "must have shafts to one side": Burlingame, *Machines That Built
America,* p. 73.

138. He sought to market: Livesay, *American Made,* p. 71.

140. Production figures: Hounshell, *American System,* pp. 155, 161.

140. "the alterations made annually": William T. Hutchinson, quoted in Hounshell, *American System*, p. 159.

140. "don't be scared man!": Hounshell, *American System*, p. 160.

141. "away from this land": Hounshell, *American System*, p. 166.

141. "the reaper works": Hounshell, *American System*, p. 154.

141. "a significant amount of handwork": Hounshell, *American System*, p. 164.

141. "has been with the Colt": Hounshell, *American System*, pp. 154, 179.

Chapter 20: Another Song-and-Dance Man

142. "Dr. Coult"; and "scientific": Burlingame, *March of the Iron Men*, p. 349.

142. "no bigger than a snuff-box": Kluger, p. 49.

142. "distinguished himself": Kluger, p. 49.

143. Colt's title as Colonel: Rosenberg, introduction to Crystal Palace book, pp. 14–17.

145. "working models": Burlingame, *March of the Iron Men*, p. 350.

145. "it was too light": Webb, p. 172.

145. "one of the most prolific inventors": Hounshell, *American System*, p. 47.

146. "I advertised in the newspapers": Webb, p. 178n.

146. "such compensation as you think": Hounshell, *American System*, p. 48.

146. "His name is almost forgotten": Burlingame, *Machines That Built America*, p. 66.

147. "did not come close": Hounshell, *American System*, p. 48.

147. "Each portion of the firearm": *United States Magazine*, quoted in Burlingame, *Machines That Built America*, p. 93.

Chapter 21: The Sewing Machine

148. "One man plods": Burlingame, *March of the Iron Men*, p. 368.

149. "Seeing this loop": Burlingame, *Machines That Built America*, p. 125.

150. "the work done on the machine": Boorstin, *Democratic Experience*, p. 93.

150. "The public are indebted to Mr. Howe": Burlingame, *March of the Iron Men*, p. 370.

151. "pig-, bar-, and sheet-iron": Charles H. Fitch, quoted in Chandler, *Visible Hand*, p. 270.

152. "the throat plate sets in": Hounshell, "The System," p. 144.

153. "A large part of our own success": Hounshell, "The System," p. 133–134.

153. "Singer probably manufactured": Hounshell, *American System*, pp. 106, 4–5.

Chapter 22: Interchangeable Parts

154. "All the separate parts travel": Colt, quoted in Howard, "Interchangeable Parts Reexamined," pp. 642–643n.

154. "where the required level": Howard, "Interchangeable Parts Reexamined," p. 633.

154. "only the federal government could have financed": Smith, "Military Entrepreneurship," p. 95.

155. "failed to produce machines": Hounshell, *American System*, p. 97.

155. "They must know how far to travel": Landes, p. 309.

155. "The micrometer made it possible": Uselding, "Measuring Techniques," p. 118.

155. "it is doubtful": Uselding, "Measuring Techniques," p. 116.

155. "in 1880 interchangeability": Mayr and Post, introduction, p. xiii.

155. "not until the twentieth century": Ferguson, "History and Historiography," p. 5.

156. "only Remington claimed": Howard, "Interchangeable Parts Revisited," p. 549.

156. Xerox story: John H. Dessauer, *My Years with Xerox: The Billions Nobody Wanted* (1971), quoted in John Brooks, ed., *Autobiography of American Business* (1974), pp. 246, 247, 248.

Chapter 23: When to Innovate, When Not To

157. "When I was at Oldsmobile": quoted in Halberstam, p. 23.

158. "inhibited by an aversion": Habakkuk, p. 112.

158. "technological development as an autonomous": Noble, *Forces of Production*, p. 217.

159. "I recollect that when my father": Sellers, p. 100.

159. For the Rockefeller story, see David Freeman Hawke, *John D.: The Founding Father of the Rockefellers* (1980).

Chapter 24: By Print and Word of Mouth

166. "at the outset": Wallace, *Rockdale,* p. 114.

166. "Nathaniel French": Smith, *Harpers Ferry,* p. 245.

167. On Franklin Institute: Sinclair, pp. 87, 94, 99, 100.

Chapter 25: An American in England: 1832

169. The information in this chapter is drawn from Sellers, principally chapters 14 through 17, pp. 108–134.

Chapter 26: Brown & Sharpe

178. "Considering their varied engagements": Smith, *Harpers Ferry,* pp. 288–289.

179. "one who thoroughly understood": Woodbury, p. 59.

179. "has taken much longer": Brown and Sharpe to Willcox, 22 March 1858, quoted in Hounshell, *American System,* p. 77.

179. "to give us business enough": Brown and Sharpe to Willcox, ca. 1 October 1858, quoted in Hounshell, *American System,* p. 78.

179. "The company reaped more": Hounshell, *American System,* p. 79.

180. "were not restricted": Woodbury, p. 60.

180. "were developed in a form": Woodbury, p. 61.

180. "It marked at once the culmination": Woodbury, p. 58.

180. "Mr. Brown's greatest achievement": Leland quoted in Woodbury, p. 67.

180. On Howe: see Hounshell, *American System,* p. 80.

181. On Leland: see Hounshell, *American System,* pp. 81–82 and Leland, passim.

182. On Norton: see Woodbury, pp. 97–108.

Chapter 27: The Franklin Institute

185. "unhampered by theory": Landes, pp. 133, 134.

185. "they lacked a publication": Wallace, *Rockdale,* p. 227.

186. "were to be found": quoted in Sinclair, p. 98.

186. "to advance the general interests": *The Saturday Evening Post*, 24 February 1824, quoted in Sinclair, p. 32n.

186. "to provide instruction to workingmen": Sinclair, p. 33.

186. "the age of secrecy in arts": Sinclair, p. 195.

187. "engaged in a large-scale cooperative effort": Sinclair, p. 90.

187. "to lay open those stores of the genius": Sinclair, p. 63.

187. "an inventive, enlightened, and inquiring": Sinclair, p. 137.

187. "in estimating the importance": Alfred North Whitehead, *The Aims of Education* (paperback ed. 1956), p. 61.

187. "for the purpose of instructing persons": Eaton quoted in Noble, *America by Design*, p. 21.

188. "that artisans will": Jones, quoted in Sinclair, p. 197.

188. "any other single person": Sinclair, pp. 149, 153.

188. "By stimulating national pride": Wallace, *Rockdale*, pp. 232–233.

Chapter 28: One More Song-and-Dance Man

190. "mechanically inept": Hindle, *Emulation*, p. 105.

191. "Well, Captain": quoted in Burlingame, *March of the Iron Men*, p. 276.

191. "which laid down the entire basis": Burlingame, *March of the Iron Men*, p. 268.

192. "By using relays": Isaac Asimov, *Understanding Physics: Light, Magnetism, and Electricity* (1966), p. 209.

192. "His compelling motive": Burlingame, *March of the Iron Men*, p. 270.

192. "not indebted" to Henry: Burlingame, *March of the Iron Men*, p. 280.

192. "readily admits": Burlingame, *March of the Iron Men*, p. 281.

192. "The primary strength": Hindle, *Emulation*, p. 93.

194. "If, on the arrival": quoted in Rosenberg, *American System*, p. 368.

Chapter 29: Water and Science

195. "has never been fix'd": quoted in Sinclair, p. 141.

196. "created an international sensation": Wallace, *Rockdale*, p. 236.

196. "the ratio of power expended": Wallace, *Rockdale,* p. 235.

196. "did much to spread": Layton, "Scientific Technology," p. 68.

196. On Tyler's wheel: Layton, "Scientific Technology," p. 67.

196. On Parker turbine: Layton, "Scientific Technology," p. 69.

197. "was one of the first instances": Struik, p. 253.

197. "developed a scientific tradition": Layton, "Scientific Technology," p. 76.

197. On Emerson's testing laboratory: Layton, "James B. Francis," p. 103.

197. On Boyden's legacy: Struik, p. 253.

198. "turbine testing into a matter": Layton, "James B. Francis," p. 103.

Chapter 30: Wood

201. "In no branch of manufacture": Joseph Whitworth, quoted in Rosenberg, *American System,* p. 343.

201. "Many works in various towns": quoted in Rosenberg, *American System,* p. 344.

202. "designed not to eliminate": Ferguson, "History and Historiography," pp. 6–7.

202. "order, system, intelligent supervision": quoted in Hounshell, *American System,* p. 147.

202. "the nation's largest carriage-makers": Chandler, *Visible Hand,* p. 248.

202. "of good quality": quoted in Hummel, p. 44.

203. "The basic new idea": Boorstin, *National Experience,* p. 150.

204. "often appallingly unattractive": Kouwenhoven, p. 52.

204. "the most important contribution": quoted in Kouwenhoven, p. 52.

204. Decline of price of nails: Rosenberg, "America's Rise," p. 43.

204. "the domestic production of nails": Robert Fogel, *Railroads and American Economic Growth,* p. 135, quoted in Rosenberg, "America's Rise," p. 198.

204. "Lumber manufacture": John Richards, *A Treatise on the Construction and Operation of Wood-Working Machines* (1872), p. 141, quoted in Rosenberg, "America's Rise," p. 202.

205. "If the saw could be reduced": Rosenberg, "America's Rise," p. 48.

206. "bare, bald white cubes": Calvert Vaux, quoted in Kouwenhoven, p. 52.

206. "ideally suited to carving curved forms": Seidler, p. 69.

Chapter 31: Coal

206. "when our wood-fuel shall": quoted in Brown, pp. 200–201.

206. On Jordan Mott: see Cowan, p. 60.

207. "there is a saving of about fifty percent": quoted in Chandler, "Anthracite Coal," p. 153.

207. On Pennsylvania Hospital and coal: Chandler, "Anthracite Coal," p. 152.

207. "and it was not long before": Cowan, p. 61.

207. "a feeling of unutterable repugnance": Strasser, p. 56.

207. "You may take the poetry": Strasser, p. 58.

208. On Solomon Willard: Boorstin, *National Experience*, p. 17.

208. On Eliphalet Nott: Albion, p. 159.

208. On gas illumination: Strasser, p. 68; Sinclair, p. 321; George T. Brown, *The Gas Light Company of Baltimore: A Study of Natural Monopoly* (1936).

209. Statistics on Pennsylvania coal production: Chandler, "Anthracite Coal," p. 158.

209. "Thereafter, in the anthracite area": Wallace, *Social Context*, pp. 122–123.

210. "It was the ventilation system": Wallace, *Social Context*, p. 150.

Chapter 32: Iron

212. "the pig iron was melted down": Morison, "John Fritz and the Three High Rail Mill," in his *From Know-How to Nowhere*, p. 78.

212. "the whole performance": Burlingame, *March of the Iron Men*, p. 109.

212. Manifests: Albion, pp. 67–68.

213. On John Fritz: Morison, "Fritz," pp. 72–86.

Chapter 33 : The Old Order Begins to Pass

214. "By the early 1850s": Chandler, "Anthracite Coal," pp. 177–178.

215. "The relations with the owners": Chandler, "Anthracite Coal," p. 178.

215. "The rushing locomotives": Mumford, *City*, p. 451.

215. "It is surrounded with a dense": Chevalier, p. 169.

215. "there is no interruption of business": Chevalier, p. 169.

216. "about thirty houses": Chevalier, p. 174.

216. "many imperfections": Chevalier, p. 174.

Chapter 34 : The Challenges Presented

219. "My good Friend": quoted in Sellers, p. 150.

219. On Lemuel Shaw: Boorstin, *National Experience*, pp. 40–41.

220. "the construction of a single mile": Hunter, *Steamboats*, p. 308, quoted in Chandler, *Visible Hand*, p. 43.

221. Camden-Amboy line: Hindle and Lubar, pp. 129–138. "There are hardly two railways": David Stevenson, *Sketch of the Civil Engineering of North America* (1838), p. 240, quoted in Sellers, p. 148n.

221. "on minutely examining": Stevenson, quoted in Fisher, *Workshops*, p. 174.

221. On gauges: Taylor, p. 82.

Chapter 35 : The Locomotive

223. "hundreds of lesser-known men": Ferguson, headnote in Sellers, p. 166.

223. On Jervis: Morison, "The Works of John Jervis," in his *From Know-How to Nowhere,* pp. 40–71.

223. "The men who did commonly design": Calhoun, p. 86.

224. "a crazy affair": Charles Barnard, "English and American Locomotoves," *Harper's Monthly Magazine,* March 1879, quoted in Kouwenhoven, pp. 31–32.

225. Treasury report on number of locomotives (1838) in United States: Calhoun, p. 86.

225. On Phineas Davis: Sellers, pp. 180–185.

227. "it is easily transported": quoted in Bishop, vol. 3, p. 43.

227. On Stephen Long: Sellers, pp. 191–192.

Chapter 36: New Founding Fathers

228. On Herman Haupt: James A. Ward, "Herman Haupt and the Development of the Pennsylvania Railroad," *Pennsylvania Magazine of History and Biography* 95 (1970): 73–97.

229. "Even the biggest of these could be explored": Livesay, *Carnegie*, p. 30.

230. "Collecting railroad revenues": Livesay, *Carnegie*, p. 32.

230. "Uniform methods of construction": Chandler, *Visible Hand*, pp. 82–83.

231. "By the coming of the Civil War": Chandler, *Visible Hand*, p. 107.

231. "Improved organization": Chandler, *Visible Hand*, p. 121.

Chapter 37: America Becomes Aware of Itself

232. "much as a lover": Chevalier, quoted in Fisher, *Workshops*, p. 72.

232. "The one seems to hear": Guillaume Tell Poussin, quoted in Fisher, *Workshops*, p. 72.

233. "vibrations emanating from New York": Leo Marx, *New York Review of Books*, 15 March 1984: 29.

233. "Then the little locomotive": Faulkner, "The Bear," (1942), quoted in Marx, *Machine in the Garden*, p. 227.

233. "a perfect symbol": Mill paraphrased in Marx, *Machine in the Garden*, p. 192.

233. Emerson quotes: Marx, *Machine in the Garden*, pp. 230, 17, 263.

233. "long shriek": Marx, *Machine in the Garden*, p. 13.

234. Thoreau quotes: Marx, *Machine in the Garden*, pp. 249, 252, 247.

234. Everett quotes: Kasson, p. 41.

Chapter 38: The Home

237. "The age of invention and mass production": Mumford, *City*, p. 465.

237. "Housework": Cowan, p. 38.

237. "of hardship": Trollope, p. 117.

238. "Get a range as plain as possible": quoted in Strasser, p. 40.

238. "Shoppers returned from the market": Strasser, p. 29.

238. "the house shrank during winter": Strasser, pp. 63–64.

239. "one wash, one boiling": Strasser, p. 105.

239. "Even the most pared-down versions": Strasser, p. 108.

239. "keeping the window open": quoted in Strasser, p. 96.

239. "their visits to the privy": Strasser, p. 96.

241. Quotations from William Gaddis: "Making Ourselves Comfortable," review of Witold Rybcznski's *Home: A Short History of an Idea*, in the *New York Times Book Review*, 3 August 1986: 23.

Chapter 39: The City

242. "transformed the industrial towns": Mumford, *City*, pp. 446–447.

242. "the dirtiest town": Chevalier, p. 169.

242. "fresh green leaf": Alexander Mackay, quoted in Fisher, *Workshops*, p. 163.

242. On Schermerhorn: See Gene Schermerhorn, *Letters to Phil: Memories of a New York Boyhood, 1848–1856* (1983).

242. "blotted the landscape": Mumford, *City*, p. 471.

243. "the old pattern of the city": Sam B. Warner, Jr., *Streetcar Suburbs* (1962), p. 17.

243. "These passenger cars": Fisher, *Philadelphia Perspective*, 1 February 1859, p. 316.

244. "The American hotel": Boorstin, *National Experience*, p. 137.

Chapter 40: The Country

245. On Russian harvest, 1906, Alex Shoumatoff, "Personal History," *The New Yorker*, 26 April 1982: 45–46.

246. "probably less than a quarter": Cochran, p. 89.

246. On hay rake: Cochran, p. 88.

246. On steam plow: Fisher, *Philadelphia Perspective*, 29 July 1859, p. 329.

246. "left off the use of common iron spades and hoes": Kouwenhoven, p. 17.

246. "as in colonial times": Giedion, pp. 144, 149.

247. "perfected [it] to a high": Giedion, p. 147.

247. "enables it to be more easily drawn out": Kouwenhoven, p. 16.

247. "The Man with the Hoe": quoted in Giedion, p. 131.

247. "The mass production of apples": *The Rural Cyclopedia* (1854), quoted in Giedion, p. 134.

Chapter 41: London, 1851

251. "dull industrial exhibition": Hobhouse, quoted in Dalzell, p. 11.

251. "Can it be possible": quoted from *New York Herald*, 26 November 1850, in Dalzell, pp. 28–29.

252. "No eagle": quoted in Sparling, p. 32.

252. "being struck by the glaring contrast": *Punch*, quoted in Sparling, p. 43.

252. "certainly not very interesting": Sparling, p. 40.

252. "We too often talk": quoted in Sparling, pp. 44–45.

252. "The absence in the United States": quoted in Rosenberg, *American System*, p. 7n.

253. "revolvers even threaten": quoted in Dalzell, p. 51.

253. "the irregular, impetuous rush": quoted in Sparling, p. 33.

253. Borden's meat biscuits: Boorstin, *Democratic Experience*, p. 311.

253. "a more simple, economical": quoted in Hindle and Lubar, p. 258.

254. "Gentlemen": Dalzell, p. 48.

254. "The reaping machine from the United States": Dalzell, p. 49.

254. "The 'Prairie Ground' is filled": Dalzell, p. 49.

254. "for years guarded the vaults": Dalzell, p. 47.

254. "our descendants on the other": Dalzell, p. 47.

254. "has created a positive furor": Dalzell, p. 49–50.

255. "Timber for Sale": Dalzell, p. 50.

255. "Yankee Doodle sent to town": Dalzell, p. 51.

Chapter 42: New York City—1853

Much of this chapter has been drawn from Hirschfeld.

256. "I was an ass": quoted in Harris, pp. 147–148.

258. "set up ratchets": Boorstin, *Democratic Experience*, p. 103.

259. "Song of the Exposition" can be found in the Bantam printing of the 1892 edition of *Leaves of Grass* (1983), pp. 158–166.

260. "a sweet sunprint": quoted in Hirschfeld, pp. 114–115.

262. "The Southern Belle": quoted in Kasson, p. 155.

Bibliography

Adams, Henry. *History of the United States During the Administrations of Thomas Jefferson and James Madison*. 9 vols. (1889–1891).

Albion, Robert Greenhalgh. *The Rise of New York Port* (1939).

Appleton, Nathan. *Introduction of the Power Loom and the Origin of Lowell* (1858), quoted in Sanford, pp. 352–359.

Ashby, Eric. *Technology and the Academics* (1959).

Bateman, Fred and Thomas Weiss. *A Deplorable Scarcity: The Failure of Industrialization in the Slave Economy* (1981).

Bedini, Silvio A. "Artisans in Wood: The Mathematical Instrument Makers," in Hindle, ed., *America's Wooden Age*, pp. 85–119.

———. *Thinkers and Tinkers: Early American Men of Science* (1975).

Beverley, Robert. *The History of the Present State of Virginia* (1705).

Bigelow, Jacob. *Elements of Technology* (1829).

Billington, David. *The Tower and the Bridge* (1983).

Bishop, J. Leander. *A History of American Manufactures from 1608 to 1860*. 3 vols. (3d ed. 1868).

Boorstin, Daniel. *The Democratic Experience* (1973).

———. *The Discoverers* (1983).

———. *The National Experience* (1965).

Braudel, Fernand. *The Structures of Everyday Life: The Limits of the Possible* (1981).

Bridgenbaugh, Carl. *The Colonial Craftsman* (paperback ed. 1961).

———. *Vexed and Troubled Englishmen, 1590–1642* (1968).

Brissot de Warville, Jacques Pierre, *New Travels in the United States, 1781*, translated by Mary Soceanu Vamos and Durand Echeverria and edited by Durand Echeverria (1964).

Brooks, John, ed. *Autobiography of American Business* (1974).

287

Brown, Ralph L. *Mirror for Americans: Likeness of the Eastern Seaboard (1810)* (1943).

Burlingame, Roger. *Engines for Democracy* (1940).

———. *Machines That Built America* (1935).

———. *March of the Iron Men: A Social History of Union Through Invention* (1938).

Cain, Louis P. and Paul Uselding, eds. *Business Enterprise and Economic Change* (1973).

Calhoun, Daniel Hovey. *The American Civil Engineer: Origin and Conflict* (1960).

Calvert, M. A. *The Mechanical Engineer in America, 1830–1910: Professional Culture in Conflict* (1967).

Chandler, Alfred D. "The American System and Modern Management," in Mayr and Post, eds., pp. 153–170.

———. "Anthracite Coal and the Beginnings of the Industrial Revolution in the United States," *Business History Review* 46 (1972): 141–181.

———. *The Visible Hand: The Managerial Revolution in American Business* (1977).

Chevalier, Michel. *Society, Manners and Politics in the United States* (1839).

Cochran, Thomas C. *Frontiers of Change: Early Industrialism in America* (1981).

Cowan, Ruth Schwartz. *More Work for Mother* (1983).

Crèvecoeur, J. Hector St. John de. *Sketches of Eighteenth Century America: More "Letters from an American Farmer,"* edited by Henri L. Bourdin, Ralph H. Gabriel, and Stanley T. Williams (1925).

Curti, Merle. "America at the World Fairs, 1851–1893," in *Probing Our Past* (1955), pp. 246–277.

Dalzell, Robert F., Jr. *American Participation in the Great Exhibition of 1851* (1960).

Daumas, Maurice. *A History of Technology and Invention: Progress Through the Ages.* Vol. 3: *The Expansion of Mechanization, 1725–1860* (1979).

Davis, Robert S., Jr. "The Machine Tools of a Southern Iron Founder: Findlay's Steam Engine Manufactory," in *Tools and Technology: The Newsletter of the American Precision Museum* 6(n.d.): 25–28.

Dew, Charles B. *Ironmaker to the Confederacy: John R. Anderson and the Tredgar Iron Works* (1966).

Dwight, Timothy. *Travels in New England*, edited by Barbara Miller Solomon. 4 vols. (1969).

Eaton, Clement. *The Growth of Southern Civilization, 1790–1860* (1961).

Ferguson, Eugene S. "The American-ness of American Technology," *Technology and Culture* 20 (1979) : 3–24.

———. "History and Historiography," in Mayr and Post, eds., pp. 1–23.

———. "On the Origin and Development of American Mechanical 'Know-How,'" reprinted in Layton, ed., *Technology and Social Change*, pp. 9–24.

———, ed. *Early Engineering Reminiscences ... of George Escol Sellers* (1965).

Fisher, Marvin. *Workshops in the Wilderness: The European Response to the American Industrialism, 1830–1860* (1967).

Fisher, Sidney George. *Philadelphia Perspective: The Diary of Sidney George Fisher*, edited by Nicholas B. Wainwright (1967).

Fitch, Charles H. "Report on the Manufactures of Interchangeable Mechanisms," in *Report on the Manufactures of the United States at the Tenth Census* (1883).

Flexner, James Thomas. *Steamboats Come True* (1944). Reprinted in paperback under the title *Inventors in Action* (1962).

Forbes, Esther. *Paul Revere, and the World He Lived In* (1942).

Franklin Papers, edited by William B. Willcox et al., Vol. 22 (1981).

Giedion, Siegfried. *Mechanization Takes Command: A Contribution to Anonymous History* (1948).

Gordon, Robert B. "Hydrological Science and the Development of Waterpower for Manufacturing," *Technology and Culture* 26 (1985) : 204–235.

Green, Constance McL. *Eli Whitney and the Birth of American Technology* (1953).

Habakkuk, H. J. *American and British Technology in the Nineteenth Century: The Search for Labour-Saving Inventions* (1967).

Halberstam, David. *The Reckoning* (1986).

Harris, Neil. *Humbug: The Art of P. T. Barnum* (1973).

Hawke, David. *A Transaction of Free Men* (1964).

Hindle, Brooke. *Emulation and Invention* (1981).

———. "Charles Willson Peale's Science and Technology," in Richardson, Hindle, and Miller, pp. 106–160.

———. *Pursuit of Science* (1956).

———. *Technology in Early America* (1966).

———, ed. *America's Wooden Age: Aspects of Its Early Technology* (1975).

———, ed. *Material Culture of the Wooden Age* (1981).

——— and Steven Lubar. *Engines of Change: The American Industrial Revolution 1790–1860* (1986).

Hirschfeld, Charles. "America on Exhibition: The New York Crystal Palace," *American Quarterly* 9 (1972): 101–116.

Hounshell, David A. "Commentary: On the Discipline of the History of American Technology," *Journal of American History* 67 (1981): 854–865, followed by an exchange with Darwin H. Stapleton, pp. 897–902.

———. *From the American System to Mass Production, 1800–1932: The Development of Manufacturing Technology in the United States* (1984).

———. "The System: Theory and Practice," in Mayr and Post, eds., pp. 127–152.

Howard, Robert A. "Interchangeable Parts Reexamined: The Private Sector of the American Arms Industry on the Eve of the Civil War," *Technology and Culture* 19 (1978): 633–649.

———. "Interchangeable Parts Revisited," *Technology and Culture* 21 (1980): 549–550.

Hughes, Jonathan. *The Vital Few: The Entrepreneur and American Economic Progress* (1973) (pb: 1986).

Hummel, Charles F. "The Business of Woodworking: 1700 to 1840," in Kebabian and Lipke, eds., pp. 43–63.

Hunter, Louis C. *A History of Industrial Power in the United States, 1780–1930.* Vol. 1: *Waterpower in the Century of the Steam Engine* (1979).

———. "The Invention of the Western Steamboat," in Layton, ed., *Technology and Social Change*, pp. 25–46.

———. *Steamboats on the Western Rivers* (1949).

———. "Waterpower in the Century of the Steam Engine," in *America's Wooden Age: Aspects of Its Early Technology*, Hindle, ed., p. 160.

Jacobs, Jane. "Why TVA Failed," *New York Review of Books*, 10 May 1984: 41–47.

Jeremy, David J. *Transatlantic Industrial Revolution: The Diffusion of Textile Technologies Between Britain and America, 1790–1830s* (1981).

Jerome, John. *Truck: On Re-Building a Worn-Out Pickup, and Other Post-Technological Adventures* (1977).

Kasson, John F. *Civilizing the Machine: Technology and Republican Values in America, 1776–1900* (1976).

Kebabian, Paul B. and William C. Lipke, eds. *Tools and Technologies: America's Wooden Age* (1979).

Klinkenborg, Verlyn. *Making Hay* (1986).

Kluger, Richard. *The Paper: The Life and Death of the New York Herald Tribune* (1986).

Kouwenhoven, John A. *Made in America: The Arts in Modern Civilization* (1948).

Kulik, Gary. "A Factory System of Wood: Cultural and Technological Change in the Building of the First Cotton Mills," in Hindle, ed., *Material Culture in the Wooden Age*, pp. 300–335.

Lacey, Robert. *Ford: The Men and the Machines* (1986).

Landes, David S. *Revolution in Time: Clocks and the Making of the Modern World* (1983).

Layton, Edwin T., Jr. "James B. Francis and the Rise of Scientific Technology," in *Technology in America*, edited by Carroll W. Pursell, Jr., pp. 92–104.

———. "Scientific Technology, 1845–1900: The Hydraulic Turbine and the Origins of American Industrial Research," *Technology and Culture* 20 (1979): 64–89.

———, ed. *Technology and Social Change in America* (1973).

Leland, Ottilie. *Master of Precision: Henry M. Leland* (1966).

Lindstrom, Diane. *Economic Development in the Philadelphia Region, 1810–1850* (1978).

———. "The Industrialization of the East, 1810–1860," in *Working Papers from the Region Economic History Research Center*, edited by Glenn Porter and William H. Mulligan, Jr. Vol. 2: *Eleutherian Mills-Hagley Foundation* (1979), pp. 17–59.

Livesay, Harold C. *American Made: Men Who Shaped the American Economy* (1980).

———. *Andrew Carnegie and the Rise of Big Business* (1975).

Lubar, Steven. *See* Hindle, Brooke, and Steven Lubar.

Martin, Edwin T. *Thomas Jefferson: Scientists* (1952).

Martineau, Harriet. *Retrospect of Western Travels* (reprint of 1836 edition, 2 vols.).

———. *Society in America*, edited, abridged, with an introductory essay by Seymour Martin Lipset (1962).

Marx, Leo. "Closely Watched Trains," review of John R. Rilgoe, *Metropolitan Corridor: Railroads and the American Scene, New York Review of Books*, 15 March 1984, pp. 28–30.

———. *The Machine in the Garden: Technology and the Pastoral Ideal in America* (1964).

Mayr, Otto, and Robert C. Post, eds. *Yankee Enterprise: The Rise of the American System of Manufactures* (1981).

Mease, James. *The Picture of Philadelphia* (1811).

Miller, Perry. *The Life of the Mind in America from the Revolution to the Civil War* (1965).

Morison, Elting E. *From Know-How to Nowhere: The Development of American Technology* (1974).

———. *Men, Machines, and Modern Times* (1966).

Mumford, Lewis. *The City in History* (1961).

———. *Technics and Civilization* (1934).

Nevins, Allan, and Jeanette Mirsky. *The World of Eli Whitney* (1952).

Newhouse, John. *A Sporty Game* (1982).

Noble, David. *America by Design* (1977).

———. *Forces of Production* (1984).

Oliver, John W. *History of American Technology* (1956).

Penn, Theodore Z. "The Slater Mill Historic Site and the Wilkinson Mill Machine Shop Exhibit," *Technology and Culture* 21 (1980) : 56–66.

Peterson, Merrill D. *Thomas Jefferson and the New Nation: A Biography* (1970).

Pirsig, Robert M. *Zen and the Art of Motorcycle Maintenance: An Inquiry into Values* (1974).

Pursell, Carroll W. "Thomas Digges and William Pearce: An Example of the Transit of Technology," *William and Mary Quarterly* 21 (1964) : 551–560.

———, ed. *Technology in America: A History of Individuals and Ideas* (1981).

Richardson, Edgar P., Brooke Hindle, and Lillian B. Miller. *Charles Willson Peale and His World* (1983).

Rolt, L. T. C. *Tools for the Job: A Short History of Machine Tools* (1965).

Rosenberg, Nathan. "America's Rise to Woodworking Leadership," in Hindle, ed., *America's Wooden Age*, pp. 37–62.

———. *Perspectives on Technology* (1976).

———. "Technological Interdependence in the American Economy," *Technology and Culture* 20 (1979) : 25–50.

———. *Technology and American Economic Growth* (1972).

———. "Why in America?" in Mayr and Post, eds., pp. 49–61.

———, ed. *The American System of Manufactures: The Report of the Committee on the Machinery of the United States, 1855* (1969).

Sanford, Charles L., ed. *Quest for America, 1810–1824* (1964).

Scanlon, Ann M. "The Building of the New York Central: A Study of the Development of the International Iron Trade," in Joseph R. Frese, S. J., and Jacob Judd, eds., *An Emerging Independent American Economy, 1815–1875* (1980), pp. 99–126.

Seidler, Jan. "Transitions in New England's Nineteenth-Century Furniture Industry: Technology and Style, 1820–1880," in *Tools and Technologies: America's Wooden Age*, Kebabian and Lipke, eds. (1979), pp. 64–79.

Sellers, George Escol. *Early Engineering Reminiscences (1815–40)*, edited by Eugene S. Ferguson (1965).

Sinclair, Bruce. *Philadelphia's Philosopher Mechanics: A History of the Franklin Institute, 1824–1865* (1974).

Smith, Merritt Roe. "Eli Whitney and the American System of Manufacturing," in Pursell, ed., *Technology in America*, pp. 45–61.

————. *Harpers Ferry Armory and the New Technology: The Challenge of Change* (1977).

————. "Military Entrepreneurship," in Mayr and Post, eds., pp. 63–102.

Sparling, Tobin Andrews. (Yale catalog of) *The Great Exhibition: A Question of Taste* (1982).

Stapleton, Darwin. "Benjamin Henry Latrobe and the Transfer of Technology," in Pursell, ed., *Technology in America*, pp. 34–44.

Strasser, Susan. *Never Done* (1982).

Struik, Dirk J. *Yankee Science in the Making* (1948).

Sturt, George. *The Wheelwright's Shop* (1923; reprinted 1963).

Tarule, Rob. "The Mortise and Tenon Frame: Tradition and Technology," in Kebabian and Lipke, eds., *Tools* pp. 28–42.

Taylor, George R. *The Transportation Revolution, 1815–1860* (1962).

Trollope, Frances. *Domestic Manners of the Americans,* edited by Donald Smalley (1960).

Uselding, Paul. "Measuring Techniques and Manufacturing Practice," in Mayr and Post, eds., *Yankee* pp. 103–126.

————. "Technical Progress at the Springfield Armory, 1820–1850," reprinted in Cain and Uselding, eds., under the title "An Early Chapter in the Evolution of American Industrial Management," pp. 51–84.

Wallace, Anthony, F. C. *Rockdale: The Growth of an American Village in the Early Industrial Revolution* (1978).

————. *The Social Context of Innovation: Bureaucrats, Families and Heroes in the Early Industrial Revolution, as Foreseen in Bacon's New Atlantis* (1982).

Webb, Walter Prescott. *The Great Plains* (1931).

Woodbury, Robert S. *Studies in the History of Machine Tools* (1972).

Index